FROM THE
SUN
TO THE
STARS

FROM THE

SUN

TO THE

STARS

JAMES B KALER

University of Illinois at Urbana-Champaign, USA

 World Scientific

NEW JERSEY · LONDON · SINGAPORE · BEIJING · SHANGHAI · HONG KONG · TAIPEI · CHENNAI · TOKYO

Published by

World Scientific Publishing Co. Pte. Ltd.

5 Toh Tuck Link, Singapore 596224

USA office: 27 Warren Street, Suite 401-402, Hackensack, NJ 07601

UK office: 57 Shelton Street, Covent Garden, London WC2H 9HE

Library of Congress Cataloging-in-Publication Data

Names: Kaler, James B., author.

Title: From the Sun to the stars / James B. Kaler, University of Illinois, Urbana, USA.

Description: Singapore ; Hackensack, NJ : World Scientific Publishing Co. Pte. Ltd., [2016] |
    Includes index.

Identifiers: LCCN 2016015331| ISBN 9789813143753 (hardcover ; alk. paper) |
    ISBN 9813143754 (hardcover ; alk. paper) | ISBN 9789813143241 (pbk ; alk. paper) |
    ISBN 981314324X (pbk ; alk. paper)

Subjects: LCSH: Sun. | Solar system. | Stars. | Planets.

Classification: LCC QB521 .K27 2016 | DDC 523--dc23

LC record available at https://lccn.loc.gov/2016015331

British Library Cataloguing-in-Publication Data

A catalogue record for this book is available from the British Library.

Desk Editor: Ng Kah Fee

Typeset by Stallion Press

Email: enquiries@stallionpress.com

Printed in Singapore by Mainland Press Pte Ltd.

To my OLLI students
and to the Osher Lifelong Learning Institute.

Contents

What It's All About

The good news is that this is not a textbook. And since we're taking in the glories of the Universe from both near and far, there is no bad news except for our increasing inability to witness them first hand given the brightening of the sky. It's instead a travel guide that begins at home with the Earth, Sun, and Solar System then, after a look at the tools of modern astronomy, tours the stellar cosmos to examine a vast variety of celestial sights. After seeing how stars are born, live, and die we end with an examination of other planets and what possibilities there might be for life. The object is to make the reader feel like Howard Carter when he looked into King Tut's tomb and said "I see wonderful things."

This work began as an unillustrated narrative to a set of eight evolving OLLI lectures that I presented at the University of Illinois from 2010 through 2014, a short version of which can be found at http://stars.astro.illinois.edu/sow/sunstar.html. I would like to thank former Director Kathleen Holden for inviting me to give the talks, current Director Chris Catanzerite for letting me continue them, the Osher Lifelong Learning Institute (OLLI) for providing the opportunity, my terrific editors Jessica Fricchione and Chad Hollingsworth, and, above all, the interested and enthusiastic older students who simply wanted to learn. Thanks and enjoy the show.

1

A Sunny Day

The Sun: A glowing ball of heat and light that everyone appreciates, especially when missing it on a cloudy day, and no one dares look at except perhaps when rising or setting in the horizon murk or when behind sufficiently thick clouds. It's just too bright. But without its intense illumination, there would be no life, as nearly all the energy we live by ultimately comes from solar power, including that derived from fossil fuels. The seemingly simple solar facade hides a deeply buried nuclear engine that has allowed the Sun to glow at close to its current luminosity for nearly five billion years. And it has another five billion to go before it begins to die away, or perhaps better put, before it transforms itself into something quite different than it is today. That is a story for later, as is its place among its billions of siblings, seen at night as all the other stars sparkling against a blackened sky.

"Just the facts..."

Said Joe Friday. If that's confusing, "You can look it up," attributed to Casey Stengel or more likely James Thurber.

The Sun looks like a solid ball. It isn't. Just the opposite, it's gaseous throughout. It does not fly apart because it's held together by its own inward-pulling gravity (Chapter 2), which compresses the Sun into a near-perfect sphere. The apparent razored edge is caused by highly opaque solar gases: you just can't look very far into the apparent solar "surface" any more than you can into a fair-weather cloud.

Figure 1.1. Though 150 million kilometers (93 million miles) away, the Sun, our star, provides nearly all the energy we need to live and flourish. J. B. Kaler.

Everything about the Sun is huge, to the point that the numbers seem to take on lives of their own. The opaque surface of the Sun, the photosphere (from the Greek, meaning "sphere of light"), glows with a temperature of 5500 degrees Celsius (almost 10,000 degrees Fahrenheit), more than double that of the filament of a 100 watt incandescent light bulb.

In the Celsius system, water at sea level (so that it's under standard atmospheric pressure) freezes at 0 degrees, boils at 100. Pull all the energy, all the heat, from a body and you hit rock bottom, absolute zero, at a temperature of $-273°C$ ($-459°F$). At this level, all atomic and molecular motion ceases (at least in principle: nature can be contrary). The negative numbers of the two temperature scales are more than a bother. It's far simpler to start the absolute Kelvin scale (after the Scottish physicist William Thomson, Lord Kelvin, 1824–1907) at absolute zero and work upward by degrees Celsius. Water then freezes at 273 K, boils at 373 K. You sit there reading at about 295 K. The solar surface temperature then becomes 5780 K. The range among even ordinary stars is spectacular, from a couple thousand Kelvin (back to the light bulb) to more than 50,000 K, with extremes to be explored that are far greater.

The Sun averages 149.6 million kilometers (93.0 million miles) away, with a 1.7 percent variation each way over the course of the year (which is *not* a cause of the seasons; we'll get to that). That's 400 times farther than the Moon, which averages 384,400 km (with a plus or minus 5.5 percent variation that influences the tides but is otherwise unnoticeable). The Sun is so far away that at freeway speeds it would take 150 years to drive there. The mean solar distance is a fundamental unit in astronomy that, not surprisingly, is called THE Astronomical Unit, or AU. The Sun as defined by the photosphere (and there is a lot outside it) appears just over half a degree across, by coincidence the same as the Moon, which makes spectacular solar eclipses possible. At the solar distance, that angle translates into a physical diameter of 1.39 million km (864,000 miles), about 1/100th of an AU, 109 times the diameter of Earth, four times the distance between the Earth and the Moon, which is as far as humans have ever traveled. Once you get to the Sun in your highly insulated car, it would take another decade to drive around it looking at the sights, which include bubbling gases, gigantic magnetic ropes and loops, immense explosions, and deep dark magnetic sinkholes with slippery slopes, many of which can dwarf the Earth. We are again in the middle of the range. Ordinary stars run from the size of our planet to nearly that of the orbit of Saturn, which has a radius of 9.5 AU. At the extreme, stars can be no bigger than a small town.

The numbers needed to describe the Sun and stars can be so large that we need a shorthand to write them out. Big numbers are usually expressed through exponents. As examples, $4 = 2^2$ (2×2, two

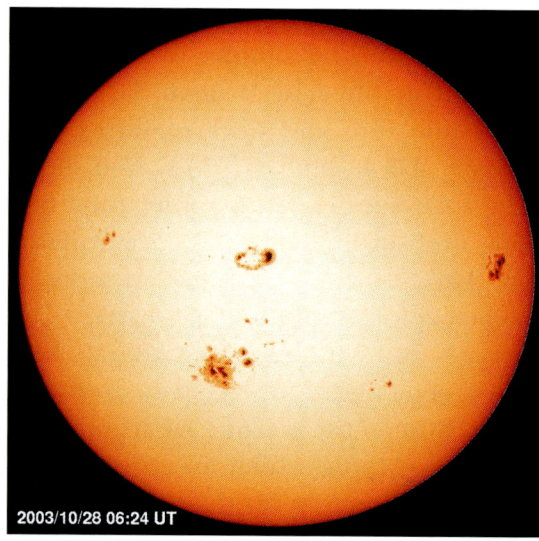

2003/10/28 06:24 UT

Figure 1.2. Highly opaque gases give the spherical Sun its sharp edge, or limb. The darkening and reddening toward the limb is the result of looking slightly more deeply, and thus to higher temperatures, at the center than near the edge. The much darker and cooler sunspots are regions of intense magnetic fields that impede the upward flow of hot gas from the interior. Solar and Heliospheric Observatory, NASA/ESA.

squared), $100 = 10^2$. A number like 10,000 is 10^4, 50,000 then being 5×10^4. Using the shorthand, on a pretend Earthly scale the Sun weighs in at 2×10^{30} kilograms (2 followed by 30 zeros), or 2×10^{27} (two thousand trillion trillion) metric tons, 330,000 times that of Earth. The Sun radiates at a rate of 4×10^{26} (four hundred trillion trillion) watts. To run the Sun for one second, you would have to pay your power company the gross domestic product of the United States for a million years. Even at its great distance, an overhead Sun delivers energy at a rate of nearly 1400 watts per square meter of ground, shining to us here on Earth half a million times brighter than the nearby Moon. Sadly, we still do not have conversion methods good enough to cover a significant part of our earthly needs. Masses of other stars range from well over 100 times that of the Sun to a lower limit that approaches the masses of our large planets, while luminosities run from millions of Suns to bulbs so dim you could not see them were they in our own Solar System.

Big numbers aside, among the most remarkable properties is the Sun's chemical composition. Very unlike Earth, and excluding the nuclear engine at the solar core, the Sun (similar to most stars) is made of 92 percent hydrogen and 8 percent helium (by numbers of atoms), which does not leave a lot of room for anything else. Within a tiny leftover bit of 0.15 percent we find all the other chemical elements, including the iron, silicon, and carbon of which the Earth is largely composed, making our planet actually quite special in spite of its small size. We are in fact a distillate of the solar gases with the light stuff missing. How do we know the solar composition? We can't get there and sample it directly (though we can come surprisingly close to doing just that). To find out, we have to peel sunlight apart.

A Sun of many colors

Sunlight looks to be a bit on the yellowish side of white. It actually shines with an amalgam of continuous colors from red through orange, yellow, green, blue, violet, and all the infinite shades in between, as demonstrated by Isaac Newton in the 17th century when he passed a narrow sunbeam through a refracting prism onto a cloth to reveal the solar spectrum. The eye then re-assembles the colors, merging them back into a visual near-pure white.

(Whatever happened to "indigo"? It's there in the blue-violet. Unless we want to get into heliotrope and mauve, there are six basic colors. But seven is a magic number. There are seven moving bodies in the sky that represent major gods: the Sun, Moon, and the five planets known since ancient times, Mercury through Saturn. Honoring them, there are seven days to the week, each named after gods, the details depending on the language. We see the "Seven Sisters" cluster of stars in the sky, even though only six are readily visible. And so on. By early logic then, there must be seven colors, so we'll make one up.)

At least in one view, light consists of a flow of electromagnetic
energy that can be visualized as electric and magnetic waves moving at
the "speed of light," in vacuum at 299,792.5 kilometers (186,282 miles)
per second. It's the speed limit of the Universe; nothing can go faster.
A light wave is described by its wavelength, the physical separation
between the wave's crests or troughs, or by its "frequency," the number
of waves that each second zip past a particular point. Multiply frequency by wavelength and you recover the speed (true of any wave,
from water to sound). There are no known limits to the wavelengths
in the electromagnetic spectrum. Light waves, optical waves if you like,

Figure 1.3. The visible solar spectrum, displayed in strips, runs from red to violet.
The blend of color gives the Sun its slightly yellowish hue. The spectrum is crossed
by hundreds of sharp dark lines. Each is produced by the absorption of sunlight
by specific atoms, ions, or even molecules, in the semi-transparent outer solar
layers. The pair just to the right of center in the third strip down is caused by
neutral sodium, while the very broad pair in the bottom strip is made by ionized
calcium. The dark line in strip six arises from ionized magnesium, while most of
the rest are produced by iron and other neutral and ionized metals. In spite of
appearances, hydrogen makes up some ninety percent of the solar gases, helium
about 10 percent, which leaves little room for heavier atoms. National Solar
Observatory, Sacramento Peak (New Mexico).

those to which the eye is sensitive, have very short wavelengths that extend roughly from 0.000040 to about 0.000075 centimeter (2.54 centimeters to the inch), less than an octave. It's more convenient to use a shorter unit, the Ångstrom, defined as a hundred millionth (10^{-8}) of a centimeter (named after the Swedish physicist Anders Ångstrom, 1814–1874). The wavelengths of the visual colors then run from 4000 Å to near 8000 Å. Waves with the shortest wavelengths (below 4500 Å) appear as violet, those longer than 6000 Å as red. In the middle, orange centers near 6000 Å, yellow around 5500, green near 5200, blue 4800, until we get back to violet, the shades varying continuously with wavelength. When passing at an angle from air or vacuum into glass or water (or any other denser transparent substance), the waves slow and bend, or "refract," toward the perpendicular, shorter violet waves more than longer red ones. Emerging, they speed up and bend again, now away from the perpendicular, which separates the colors even more: and we see Newton's spectrum.

Up in the air

Whether aware or not, everyone is familiar with the solar spectrum, as it plays out in a great variety of attractive ways, most notably in the rainbow. An afternoon thunderstorm blows off to the east, allowing sunlight to fall upon its raindrops, each of which acts as a small Newtonian prism. Single one out. As a sunbeam enters the droplet, it bends and splits into its component colors, red through violet. Reflecting from the drop's backside, the spread beam exits and widens yet more, the reversal in direction sending the split ray back into your eye. The combination of all drops combined with the refractive properties of water creates a circular colored ring 42 degrees in radius around the point below the horizon opposite the Sun. Since its waves refract the least, red will be on the rainbow's outside, blue or violet (because they refract more) on the inside. Sunbeams that hit the droplets just right can reflect twice off the backside, which produces a second, larger bow outside the bright one of 51° radius, the double reflection also reversing the colors. If the inner bow is at all bright, the outer one will always be there.

Sometimes you'll see a series of pink "supernumerary" bows inside the main one that are caused by light waves interfering with one another; if two waves fall wave-to-wave, they add together, while if they fall wave-to-trough, they cancel, giving us a bright-dark-bright-etc. pattern. If the Sun is too high, more than halfway up the sky, the point opposite the Sun will be below the horizon and the main rainbow cannot be seen against the sky. Rainbows thus appear in the early morning or late afternoon when the Sun is low enough to loft the bow into visibility.

Separation of solar colors is a natural part of the day. "Why is the sky blue?" asks the curious child. An alternative view of light is that it's made of speeding massless particles, or photons. The modern concept in the weird world of quantum mechanics (which considers the realm of the very small) is that light is both a wave and a particle at the same time. Its behavior depends on how you observe it. You might think of a photon as a chunk of a wave, though that's not a very accurate description; nor is any other. Some of the incoming solar photons bounce off air molecules (made mostly of nitrogen and oxygen), which changes the photons' directions. The process is far more efficient at shorter wavelengths that are closer to the molecules' sizes. Violet photons (they are not colored, that is just the effect they have on the eye) are fiercely scattered, whereas longer red photons are not affected much at all. Blue and violet solar photons then get knocked all over the sky, resulting in some coming into your eye from any direction. But there are relatively few violet photons in sunlight, nor is the eye very sensitive to them, so most of the scattered photons we see are blue. Hence the glorious azure sky, which varies in its shade according to the angle above the horizon and angle to the Sun. But it's all still sunlight. The blue sky is so bright that the stars disappear behind its veil. Only Venus glimmers through it.

If you pull blue and violet photons out of sunlight, the Sun itself must appear to be colored more toward the longer wave end of the spectrum. The complement to a blue sky is then a somewhat yellowed Sun that turns golden as it approaches setting, and at the extreme turns reddish. The sky looks as if it is a dome over your head. It isn't. The air is actually a thin layer that hugs the ground and fits to the curvature of the Earth, the density and pressure dropping off quickly with height, as anyone who climbs mountains will happily tell you. Toward overhead, you see outside to space through the thinnest part of the layer. As your gaze drops downward, you have to look through more and more air. A third of the way up from the horizon, you see through double the overhead thickness, while at the horizon itself you look through 38 times more air than when you look directly above you. As a result, the light of the setting Sun has a longer pathway over which it is scattered, and with more blue light removed, the Sun turns both dimmer and redder. The effect is enhanced by the absorption and scattering of solar photons not just by the air, but also by watery haze and pollution, both natural (from volcanos) and artificial. Light clouds near the horizon will reflect the reddened sunlight, giving us wonderful orange and red sunsets. The effect increases during early twilight after the Sun falls below the horizon but can still illuminate the upper atmosphere, for a brief time (until the intervening Earth gets in the way) giving sunbeam even more air to contend with.

Just as liquid water refracts and disperses the colors, so does ice. Ice crystallizes into six-sided hexagons (remember the snowflake on your mitten?) that form tiny prisms. Put thin icy clouds in front of the Sun, and they can create a colored ring 22 degrees (the bending angle through the sides of the hexagon) in radius with the Sun in the middle. Because the bending angle increases with decreasing wavelength, the "22-degree halo" is red on the inside and blue on the outside. (Be sure to hide the Sun behind something so as not to look at it directly.) As the Sun sets, spots on the ring on a line through the Sun parallel to the ground grow in intensity until these sundogs (mock suns, or parhelia) dominate. A white ring parallel to the ground caused by icy reflection may in rare instances extend through the sundogs and encircle the entire sky. Refraction through the square sides of the icy prisms produces a rare ring 47 degrees in radius. Various arcs tangential to the rings might paint the blue canvas as well. Most are related to the 22-degree ring, though one, the "circumhorizontal arc," which is attached to the lower edge of the

47-degree ring, can be bright and beautiful even when the ring itself is not there, leaving a sort of rainbow floating in the sky beneath the Sun. And if you live in the wintery north and don't want to wait for a solar halo to see refracted solar colors, just look at the sparkles of sunlight off crystals of new snow.

Great fun can be had by looking out an airplane window. Here we might see especially intense sundogs, as well as the pure reflection of the Sun off light icy clouds below the craft, making a white "subsun" that follows along with you, the clouds sometimes so transparent as to be practically invisible. They've been taken for UFOs (see Chapter 8). If you are positioned right, you might see the shadow of the plane cast on the clouds below. Interfering light waves combined with reflection may produce a colorful halo, the "glory" (sometimes called the pilot's rainbow), around it. At best, multiple colored rings grace the cloud deck, and if the clouds are far away, the airplane's shadow becomes lost, leaving nested colored rings floating in the air.

Clear air not only scatters, but as a "substance" (albeit a light one), it must also refract. As sunlight enters the layer of air above us, it bends downward, toward the perpendicular. All celestial objects as viewed from the Earth's surface are therefore lofted upward, appearing higher in the sky than they really are. As we look toward the horizon, where the angle between incoming sunlight and the air layer decreases and the path length through which we look to see the Sun increases, the effect is powerfully magnified. At the horizon, the Moon and Sun are lofted upward by half a degree, which is the angular diameter of the lunar and solar disks. See the Sun (when safe, through haze or clouds) with its lower edge (or "limb") sitting on the flatlands of the prairie or on the distant ocean's surface. It's not really there. If you could magically remove the Earth's air, the Sun would go down by half a degree and would drop from sight! But there it is, compliments of refraction. The effect extends daylight hours by several minutes, the amount depending on latitude.

Since it has to go through the air at a lower angle, the bottom of the rising or setting solar disk is refracted upward more than the top, which squashes the Sun into an oval. Since the lower part of the rising or setting Sun must pass also through more air, it will have more short waves removed and will be redder than the upper part. The Moon behaves the same way, but unless near its full phase is harder to see when just rising or setting.

Along with refraction must go the dispersion of colors. Think of the "white" Sun as consisting of overlapping disks of the simplified spectral colors, red through violet. Upward refraction will slightly separate the disks, putting violet on top, red on the bottom. But the violet and blue disks are scattered out, leaving the almost-set Sun with a brilliant green upper rim. As the last thing to disappear over a flat horizon (and seen best over the ocean), the vivid green rim produces the "green flash" so beloved by bar loungers in Key West. Though you cannot see the effect with the naked eye, stars behave the same way. Through the telescope or even powerful binoculars, stars near the horizon are stretched out into tiny spectra, with red on the bottom, green to violet on top depending on the state of the air and the angular altitude above the horizon. The result is quite pretty, but is noxious for astronomers trying to acquire stellar spectra, as the starlight enters the telescope's spectrograph (the device that produces and records celestial spectra, see Chapter 3) "pre-dispersed," an effect that must be accounted for.

Blend the colors back to the yellow-white glow of pure sunlight. As the Sun rises or sets across a lake or ocean, it throws forward a "glitter path," making it look as if you could walk into space. It's caused by reflection of sunlight on the irregular waves. Lunar glitter paths are featured in romantic movies. Even bright planets make them. Tip the glitter path upside down, where the "ocean" is made of atmospheric ice crystals that form light clouds, the same ones that can make solar halos and sundogs. Now it seems to rise upward into the sky as a "sun pillar" or its lovely cognate, a "moon pillar."

Icy or watery hazes can be found nearly everywhere. On land they are complemented by dust blown upward from the ground by winds. The dust motes, water vapor, and icy crystals are lit by reflected sunlight. A cloud in front of the Sun casts shadows. The effect is that sunrays seem to project through holes in the cloud or are cast into the sky from the cloud's ragged edges. The Sun is so distant that the sunbeams are actually parallel to one another. But from the ground, perspective makes them seem to converge back to the solar disk, and if conditions are right to create a glorious "sunburst" effect that appears to be thrown by Apollo himself. With the drab and inaccurate name of "crepuscular radiation," the phenomenon is especially pretty when seen at or even before sunrise when the air is moist and the rays seem to climb from Earth to Heaven. You'll see it a lot in grade-school art. Add to it a hidden Sun shining through the thin layer at the cloud's edge to create a "silver lining" and the results are ... celestial. When the sunrays (really the intervening cloud shadows) project to the ground, folk philosophy says that the Sun is "drawing water" back into the sky. To the contrary, it's mostly the water already in the air that makes the effect possible. If the Sun is below the horizon in morning or evening twilight, sunlight shining between the gaps between distant mountains that are themselves over the horizon can do the same thing.

Shadows are everywhere. What is night, but our being in the shadow, not of clouds, but of the Earth. And you can watch it grow. As the Sun drops below the western horizon, the Earth's shadow rises as a gray band in the east that climbs higher until it sweeps over the entire sky like a dusky hood. Night does not fall; it rises. Above the band, the upper atmosphere still catches some sunlight and scatters it downward, giving us a peaceful period of twilight that dims as the stars begin to come out. You'll see night "fall" in the west during dawn, and then at the moment the first glimmer of sunlight splashes over the horizon, the grey band sets. And the stars must wait for another night. Extend the concept of shadows farther into space and we encounter our planetary partner, the Moon.

By the silvery Moon

Sunlight also plays with the Moon, which orbits the Earth in just under a month and shines by reflected sunlight. Like the Earth it has both daytime and nighttime sides. The phases are caused by our seeing varying portions of the daytime lunar hemisphere as the

Figure 1.4. The Moon shines only by reflected sunlight. As it orbits the Earth it reveals growing then diminishing portions of its sunlit hemisphere. Here it's in the waxing (growing) gibbous phase, with the sunrise line running down the left hand side. The lunar surface is pocked with craters caused by impacts from bodies swept up during the late stages of planet formation more than four billion years ago. The smoother dark regions, or maria, are solidified lava flows, some of which fill huge impact basins that were formed well after the era of heavy cratering. Mark Killion.

angle between the Sun and the Earth-encircling Moon changes over a 29.5 day cycle, that is, by our seeing different portions of lunar daytime versus the amount of shaded lunar night. They are not caused by the shadow of the Earth.

The sky is illusory. Look upward and all celestial objects appear at the same distance. They aren't. The sky actually has three dimensions: two on the apparent sphere plus depth, which is insensible. To follow the phases, keep in mind that the Sun is actually 400 times farther than the Moon. Moreover, the Moon is a three-dimensional sphere, not a flat disk. When the Moon is between us and the Sun, we are presented with lunar night, and the Moon, called "new," is invisible. As the Moon orbits, just after the new phase, we get our first glimpse of lunar daylight, the Moon appearing night after night as a "waxing" (growing) crescent in western twilight. When the Moon and Sun are at right angles to each other, half the sunlit face shines down on us, while the other half is still in night. Since the Moon has gone through a quarter of its orbit, the "half Moon" is also called the "first quarter;" welcome to the strange world of astronomical nomenclature. As we see more and more of the sunlit hemisphere, the Moon fattens into a "gibbous" shape, and when it is finally opposite the Sun, all of the daytime side is presented and the Moon is "full." The apparent hugeness of the rising full Moon is also illusory, as it's always the same half a degree in diameter. The slight variation in angular size caused by the somewhat out-of-round orbit is not noticeable. The phases then just repeat themselves backwards while facing in the other direction, the bright side of the Moon always pointing toward the Sun. The full phase quickly passes into the waning gibbous, and after three quarters of the orbit has been turned since new, the Moon hits its eponymous third quarter, which is followed by a waning crescent that slims until the orbit is completed and we are back to new.

"If you can see the Moon in daytime, how come you can't see the Sun at night...?" goes the classic bad astronomy joke. The phases are tied to the Moon's visibility cycle. The waxing crescent has not turned very far past the Sun and is thus visible only in early evening as it chases the Sun toward setting. Being 90 degrees east of the Sun, the first quarter rises around noon and is visible during the afternoon and the first half of night, setting around midnight. Since it's opposite the Sun, the full Moon rises at sunset and sets at sunrise. With a quarter of the orbit left to go until new, third quarter rises around midnight and is visible in the morning daytime hours. The waning crescent then rises just before sunrise. Note that no matter what the phase, we see the same features all the time. The Moon rotates in synchrony with its orbit, and thus keeps the same side facing us, the result of tides raised by the Earth (Chapter 2).

The lunar orbit is tilted by about five degrees to the orbit of the Earth. As a result, new Moon usually passes above or below the Sun.

When the alignment is more or less straight, which happens twice or more per year, the new Moon at least partially covers the Sun and we may see a solar eclipse. Expressed another way, we are hit by the shadow of the Moon, which is at best tiny, so that only a small track on the Earth gets to experience a total eclipse wherein the solar disk is completely covered, allowing the faint but lovely outer solar corona (examined later) to be seen and studied. Half the time, the Moon in its non-circular orbit is too far for complete solar coverage, leaving a ring of sunlight — an annulus — around the darkened Moon, producing a scientifically uninteresting "annular eclipse."

The Earth's shadow that rises to make the night extends as a slimming cone far out into space, three times the distance of the Moon, and is big enough for the Moon to be completely immersed within its embrace. Still, the full Moon usually passes above or below it. But about twice a year, the Moon passes through the terrestrial shadow and we now might admire a lunar eclipse that is visible from an entire earthly hemisphere. Yet even in total eclipse, the Moon glows a bit as a result of sunlight scattered and refracted by the Earth's atmosphere into our planet's shadow. The lunar visibility in total eclipse depends on the clarity of the air, particularly on recent volcanic activity that may have spewed particles high aloft.

As seen from the Moon, the Earth goes through the same phases, just in reverse. When the Moon is new, the Earth as seen from the Moon must be full. When the Moon is in its early waxing crescent phase, a visiting astronaut would see a large, highly reflective gibbous Earth that is so bright as to dazzle the lunar landscape. On Earth we can view the whole ghostly lunar outline, the Moon's nighttime side aglow with "earthlight." We see the same thing in the morning near dawn when the Moon approaches new once again and our "lunarian" sees the Earth approaching full phase.

Since it's really just sunlight, moonlight causes the same atmospheric phenomena as does the Sun. From the dark countryside, under full moonlight the sky glows a faint but distinct blue. Light icy clouds can make the 22-degree halo along with moondogs and various other accompaniments, while falling rain can create ethereal moonbows. Among the more familiar spectral phenomena is a tight colored ring contiguous with the lunar disk caused by interference among the wave-like photons as they whiz by a transparent cloud's water or ice particles on their way down. The colors are reversed from that of the large halo, with blue now on the inside, red on the outside. You see the same thing through foggy glasses around bright lights. Such rings are actually more common around the Sun, but are dangerous to look at because of the Sun's brightness, and properly go unheeded. And are there few more pleasing sights than a bright Moon tempting us with its silvery glitter path across the sea, asking us to climb aboard as we further pursue the spectrum and its glories.

Expanding the spectrum

Look to the red end of Newton's rainbow, which deepens in color until it seems to evaporate into the air. "Is that all there is?" Does the spectrum stop, or do we just not see it? William Herschel found out in 1802 when he placed a thermometer off the red end of the solar spectrum and watched the temperature rise. He had discovered solar radiation more red then red, the *infrared*. Infrared's longer wavelengths just do not stimulate the eye.

Herschel (1738–1822), a founder of modern astronomy, was a Hannoverian musician who took up residence in England in 1757. Highly proficient, his compositions are recorded and heard yet today. Fascinated by astronomy, he built the finest telescopes of his era. With them, he discovered the planet Uranus and, with his sister Caroline (1750–1848), who followed him to England in 1772, catalogued thousands of celestial objects. The infrared (IR) extends from off the red toward much longer wavelengths until, at a wavelength of a millimeter or so, we arbitrarily call the radiation "radio," a part of the electromagnetic spectrum that by the early twentieth century was being vigorously explored and exploited for both science and commerce. With wavelengths that extend to kilometers long, the radio spectrum has no known outer limit.

Given Herschel's discovery, one might expect the deep violet to metamorphose into something else as well. Sure enough, shorter than violet lies the ultraviolet (UV), to which our eyes are again insensitive (but our skin is not; see below). At wavelengths a few hundred times shorter than violet, we call the electromagnetic radiation "X-rays." Much shorter yet and they are labelled "gamma rays," to which again there is no known short end. (In the early days of studies of radioactivity, physicists found three kinds of "radiation" that they called alpha, beta, and gamma. Alpha and beta "rays" are respectively helium nuclei and electrons, while gamma rays are real electromagnetic radiation; again see below). The lines of demarcation between the various kinds of radiation overlap; they are neither firm nor really defined. Whatever the wavelength, they are all the same thing and their photons all move in vacuum at the speed of light.

Electromagnetic radiation is the chief vehicle for transmitting energy throughout the Universe. The energy carried by a photon, be it X-ray, yellow light, or 10-km radio, depends directly upon its frequency, that is on the number of vibrations passing by per second, hence inversely on its wavelength. Visual photons, to which our air is mostly transparent, are energetic enough to heat the Earth and excite the chemistry of the eye, while longer infrared ones are more benign and are commonly felt as heat. Quite long, radio waves are almost totally harmless (a distinct exception being those used in a microwave oven). At the other end, the ultraviolet can be quite damaging, while mis-used X and gamma rays are increasingly deadly. ("Cosmic rays" are sometimes given status as more energetic than gamma rays. Indeed they can be, but then they are not electromagnetic photons. They are instead true particles, atomic nuclei, mostly protons, sent to us from exploding stars. We'll be struck by them later.)

If placed in a colder environment, a body above absolute zero ($-273°C$, $-459°F$) must radiate its internal energy away. It can be a solid or even made of a dense opaque gas (like the Sun). If really cold, it has little internal energy and all it can release is low-energy radio waves. Warm the thing up to a few tens of degrees above absolute zero, however, and infrared radiation pours out. The Sun, at nearly 6000 Kelvin, produces copiously in the energetic visual spectrum.

Hotter stars, in the tens of thousands of Kelvins, radiate much of their energy in the ultraviolet, and many are the ultrahot, X-ray, even gamma-ray, sources. No matter where in the spectrum a heated body achieves its maximum radiative power, however, it also shines well at all the lower energies. The Sun, for example, sends us infrared and radio (plus quite a lot of UV from magnetic phenomena). The effect of temperature is subtly seen through star colors. Cool stars appear reddish. As temperature rises we march through the spectrum toward shorter and more energetic wavelengths, from red to orange, yellow, white (substituting for green), then blue. Especially cool stars radiate only in the infrared.

It helps to define a kind of ideal radiator called a "blackbody." Being "black," it absorbs all radiation that hits it, reflecting none. To maintain its internal energy as measured through its temperature, a blackbody must emit just as much energy as it absorbs. It can therefore perversely be quite bright. The beauty of the concept is that blackbody radiation is amenable to mathematical description and allows precise formulae for how radiation intensity changes with wavelength and temperature, thus allowing temperature measurement. The "Wien law" (after the German physicist Wilhelm Wien, 1864–1928) says that the wavelength of peak energy production is in inverse proportion to the temperature. Double the temperature and the wavelength of maximum radiation is cut in half. The powerful "Stefan–Boltzmann" law states that the amount of radiation emitted per unit area (square foot, meter, furlong) of a blackbody's surface is proportional to the fourth power of the temperature. Double T and 16 times as much energy pours out. (Jozef Stefan, 1805–1903, was a Slovenian cum Austrian physicist; Ludwig Boltzmann, b. 1844, also an Austrian physicist, died tragically in 1906.) Both these rules were discovered in the laboratory and were later replicated by quantum theory, the theory of the very small that includes the wave–particle duality of light. While stars and other dense celestial objects are not true blackbodies, they can come close, close enough for the theoretical formulae to be effective in explaining what we see. The rules show on an atomic level with hard numbers why hot stars are bluer than cool ones, and for the same radius are vastly brighter.

Far from being of only academic interest, the extension of the electromagnetic spectrum frequently makes news, even leads to sharp political discussions. While our protective atmosphere is mostly opaque to ultraviolet light thanks to absorption by a high layer of ozone (a form of molecular oxygen), a little bit just off the violet sneaks through to the ground to cause sunburns (hence tans), which get worse the higher the Sun is in the sky thanks to less air in the way. We are dependent on the ozone, whose amount is sensitive to atmospheric pollution. On the academic side, this "near UV" is about all the astronomers get of high energy radiation without going into space.

The infrared is not blocked continuously, but by a variety of wavelength bands, allowing some lower-energy radiation to penetrate to the ground such that we can still see to the outside and practice infrared astronomy. The Earth, heated by sunlight to around 300 Kelvin, radiates only in the infrared and radio domains. With low-energy radiation absorbed in the infrared by water vapor, carbon dioxide, methane, and other atmospheric ingredients, the air acts as a blanket that helps keep in the heat, a phenomenon called the "greenhouse effect." It's necessary for life; without it, the Earth would be too cold to have sustained life or even to have birthed it. Here is the foundation of the argument for human effects on global warming, as increasing the carbon dioxide content through the burning of fossil fuels should act to increase Earth's average temperature and change climate patterns, observation clearly supporting theory.

Why is the air transparent to some wavelength bands and not to others? To find out, we wander down another path to look at atoms and the molecules built from them.

The very small explains the very large

The chemical elements are the 90 or so basic substances, the different kinds of atoms, out of which everything is made, and include such familiar things as hydrogen, helium, carbon, iron, gold, and uranium. But they are not, as Democritus thought, indivisible.

Rather they are assembled from three different kinds of particles that are related to the electromagnetic force (and thus to photons), which, unlike the familiar attractive gravitational force, has two sides. The proton, with a ridiculously small mass of 10^{-24} gram and diameter of 10^{-13} centimeter, carries the positive electric charge, while the electron carries the negative, even though it has just 1/1800th of the proton's mass. The electron has no known radius, if that term has any meaning here at all. In between is the neutral neutron, which has the proton's mass, but is without charge (thus fostering another bad joke).

The positive protons and neutral neutrons are combined in varied ways to make an atom's nucleus. One proton makes hydrogen, two helium, while six gives us carbon, 26 iron, and so on up to the last "natural" element, uranium, which has 92. The number of attached neutrons starts at zero for ordinary hydrogen then as we progress increases a bit more rapidly than the number of protons. Though the protons are repelled by their similar electric charge, they are stuck together by yet another force, the powerful short-range "strong force," as are the neutrons, which help glue a nucleus together. To complete the concept, the atomic nucleus may be surrounded by a cloud of "orbiting" negative electrons equal in number to that of protons to keep the atom electrically neutral. The various elements are stacked in the chemist's famed periodic table according to the ways in which they can be locked together to make an almost infinite variety of molecules, which in turn make our world and our lives possible. The scale of the atom is underappreciated. Neutral hydrogen's lone electron circles its proton at a distance of about an Ångstrom (10^{-8} cm), 100,000 times the proton's size. The atom is thus mostly filled with what we would call "empty space." But it isn't really. What you feel as solidity is not the matter itself, but the pervading electromagnetic force that prevents you from walking through a closed door.

There are two variations on the theme. For any given chemical element, or kind of atom, the number of attached neutrons can vary to make different isotopes. Normal hydrogen has just one proton and no neutrons, so it's H-1. Stick on a neutron and you have two particles, which makes it H-2 (deuterium). But there are limits. If a nucleus has too many or too few neutrons for its liking, it falls apart, whence it can emit dangerous particles (perhaps alpha and beta "rays") and high energy gamma radiation. Though for any element one isotope usually dominates, most kinds have a range of stable ones. But some have none at all, including technetium (number 43), promethium (61), and all the isotopes of elements heavier than bismuth, which has 84 protons. The best known examples are probably radium-226 (88 protons, 138 neutrons) and uranium-238 (92 and 146), which is feared when it is enriched with U-235 (92 and 143) to be used in nuclear power plants or atomic bombs. Some, like U-238, are long lived and are found readily in nature. The second thematic variation involves electrons, which are removable (even addable). Take an electron from a neutral atom and you get a positively charged "ion." One removed from helium yields singly ionized helium (He^+), while dumping two gets the doubly ionized form, He^{+2}, a bare nucleus also called an "alpha particle," as presaged above. Hydrogen with two electrons, H^- (H-minus), is responsible for the opaqueness of the solar photosphere.

Atoms of all elements, as well as their ions, both absorb and emit radiation in the form of photons, the job usually being handled by the electrons, which can change their orbital energy states (increasing with photon absorption, decreasing with photon emission). With some distinctive exceptions, astronomers, especially those in the realms of stars and interstellar clouds, deal with gases. But (and here is the key) gases at low pressure absorb and emit only at particular energies that are peculiar to the atom or ion and the particular allowed energy states of its electrons. In the case of a hot very high-pressure gas, or that of a solid, all the particular allowed energies smear together to make a pure unbroken, or continuous, spectrum.

Put a lower-pressure gas in front of a glowing higher temperature dense body or incandescent solid, and the rarefied one will absorb specific very narrow bands of color that produce something of a barcode of dark "absorption lines" set against a bright background. Each of these relates to some change in the orbiting electrons' energies. The appearance of the absorption spectrum depends on the element or ion that does the absorption, each kind of atom or ion producing a unique pattern. That of hydrogen is remarkably simple. In the optical spectrum consisting of just four narrow lines, or gaps (one in the red, another in the blue-green, and two in the violet, but many more in the IR and UV), the pattern is instantly recognizable. If you see the "fab four," you have hydrogen, whether here or across the Universe. Then, with an increasing number of electrons, the patterns rapidly become very complex. Helium has dozens of absorptions, iron millions. Observation is handled by a "spectrograph" that spreads the light into its spectrum. Older ones used Newtonian prisms. Modern ones employ "gratings" similar to the grooves on a compact disk. They work through the interference of light waves to produce a spectrum like the tight ring around the Moon, but one far more extended and pure. If we use the spectrograph to look at the low pressure gas by itself, without the background, the observer sees the reverse, bright "emission lines" at the same wavelengths as seen for the absorption lines (more on this the subject in Chapter 5).

To a rough approximation, the outer layers of the Sun that make up the partially transparent photosphere act the same way. As the continuous unbroken spectrum of the higher pressure gases of the lower levels passes through the lower pressure upper layers, the latter superimpose the absorption barcodes of all the elements, which overlap each other in a huge grand mess as seen in Figure 1.3. Using pure laboratory spectra and a lot of work, astronomers can disentangle and identify the absorptions with particular elements or their ions.

The strengths of the absorptions, that is the amounts of energy removed by the narrow bands, coupled with the theory of atomic structure (including how each absorption line is created) allow us to learn all manner of solar properties, including the solar temperature, gas density, and the all-important chemical composition, which by numbers is 92 percent hydrogen and 8 percent helium (the helium content actually found from other means). The slim 0.15 percent that remains is topped by oxygen, which is followed by neon, carbon, then nitrogen. Nearly all the elements of the periodic table have been identified in the Sun. For the most part, as the number of nuclear protons increases, the abundance relative to hydrogen declines, though there is a substantial peak around iron. Those elements that have not been found simply have absorptions too weak to be detected or are hopelessly radioactive. (We can get the abundances of most from primitive meteorites, ancient asteroids that get through the Earth's atmosphere.) The compositions of most other stars are similar. Of great fascination are the ones that aren't.

How the solar chemical composition changes with increasing proton number is one of the great data sets in astronomy. Almost all the elements in the Universe heavier than helium were forged in stars, including those in the Sun (and in us). Theories of element creation and delivery mechanisms into the interstellar clouds that birthed the Sun and other stars are tested by trying to replicate the solar chemical patterns. Given the complexity of the tasks, they do so remarkably well, as we will see when we look at dying stars, at their winds and explosions.

Molecules, combinations of atoms, produce extraordinarily rich sets of absorptions, whole bands of them, which refer us back to the absorption of infrared and ultraviolet radiation by Earth's molecular atmosphere, to ozone and the greenhouse effect.

Inside the Sun

The examination of the Sun's interior is a daunting task, as we can see directly no more than perhaps a thousand kilometers into the solar gases. We get a bit of help from "limb darkening," that the Sun is dimmer and redder at the edge (the "limb") than it is at the center (Figure 1.2). At the limb, we look into the foggy photosphere at a steeper angle than the perpendicular, so our vision does not penetrate as deeply as it does when we look at the center of the apparent solar disk. The variation in brightness away from the center combined with the radiation laws shows the rate at which temperature increases with depth. Theory, which agrees, can then penetrate to the solar core on its own.

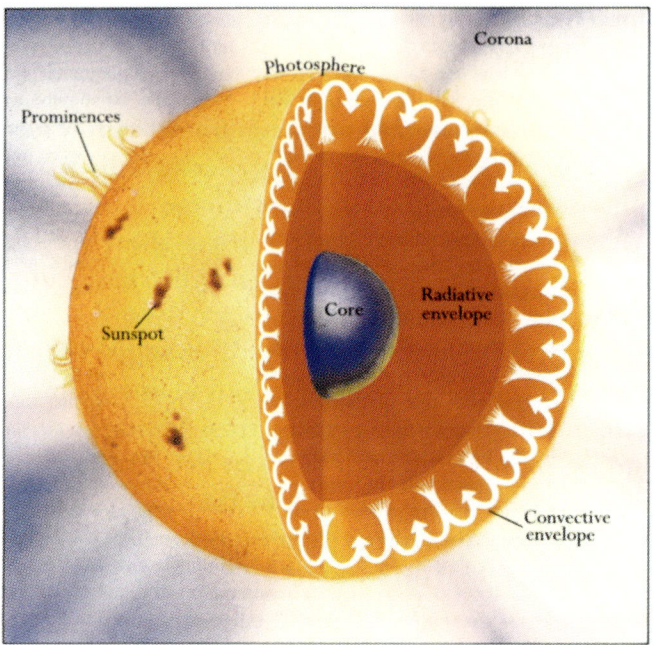

Figure 1.5. The outer third of the Sun is in a state of roiling convection in which hot gases float upward, then cool and fall back in. Beneath the convective layer is a quiet envelope through which energy is transported by radiation. The core is so hot (16 million Kelvin at the center) and dense that hydrogen is fused to helium, which generates solar energy and keeps the Sun from collapsing under its own weight. It has not changed much for 4.6 billion years and has another five or so left to go before it begins to die. From "Stars," J. B. Kaler, Scientific American Library, Freeman, New York, 1992.

To make further sense of the solar innards, look more closely at the outside. The seemingly solid opaque gaseous surface, the photosphere (which creates the remarkably complex solar spectrum of Figure 1.3) is broken up into Texas-sized bright "granules" surrounded by dark lanes, all of which come and go over periods of minutes. The granules are the tops of giant up-and-down convection cells (hot gases expanding and rising, cool ones falling) that carry heat and light being passed up from far below. The turbulence plus giant magnetic disturbances cause the Sun to ring like a bell with huge numbers of overtones, from which we can do seismic studies to peek inside much as we do for the Earth with vibrations from earthquakes. The convection layer is seen to extend a third of the way in, agreeing with what we expect from a gaseous ball of a solar mass compressed together by its own gravity. The oscillations are also sensitive to helium abundance, which provides one route to its measure.

Farther down, the immensely heated gas is quiet and moves energy outward by successive atomic absorption and re-emission of photon radiation. At the heart of the Sun is the seat of solar energy, the solar core. Occupying about a quarter of the solar size and half the solar mass, the core is calculated to be so hot (16 million Kelvin at the center) and dense (a dozen times the density of lead, but still a gas) that lighter elements can be transformed into heavier ones, in the Sun's case, four hydrogen atoms into one of helium. In the conversion, a little mass (M), 0.7 percent, is lost. But it can't just go away. Instead it's changed into energy (E) via Einstein's renowned equation, $E = Mc^2$, where "c" is the speed of light, which when squared converts a small amount of mass into a whopping amount of energy.

Here is the source and keeper of all life. If the Sun were to work off just gravitational energy, staying hot while squeezing down, it could shine for no more than 10,000 years. But the fossil record and particularly the record of radioactive decay, say otherwise. The uranium-238 atom decays largely by spitting out helium (He-4) nuclei, alpha particles. All the helium on Earth has come from this process. (Think about that at your next birthday party.) As the U-238 nucleus falls apart, it turns itself into various other nuclei that include radium and the infamous radon (do you have a detector?). The chain finally ends at lead-206, a lighter version of the lead-208 we are generally familiar with in stage weights, fishing sinkers, and lead shot. Decay rates are measured through an isotope's "half-life," the time it takes to cut the amount of a given sample in half. That of U-238 is 4.5 billion years. From the half-life and the lead-206/U-238 ratio in the oldest rocks we can find (meteorites from the asteroid belt) combined with other similar isotope ratios, we get an age for the Solar System, hence that of the Sun (all of it born together), of 4.6 billion years. Combination with the fossil record reveals that the

Sun has been shining at something close to its present rate for nearly all that time.

The only sufficient energy source we know, one that explains the solar energy and lifetime, is the nuclear fusion of light elements, the opposite of the radioactive fission of the heavies. By analogy, all stars work, or have worked, this way. In the broad picture you get energy by fusing elements upward in the periodic table to iron and by fissioning elements down to iron (with a number of exceptions), which explains why iron is so common. More specifically, helium is made from hydrogen by the proton–proton (p–p) chain. At the core's temperature and density, protons can slam into each other at speeds so great they can overcome the electrical repulsion between positive charges and get close enough together so that they stick under the short-range strong force, producing heat in the form of a gamma ray. In the process, one proton turns into a neutron with the ejection of positive electron (anti-matter!) and a strange particle called a "neutrino." The result is heavy hydrogen, H-2, or deuterium. The "positron" then hits an ordinary negative electron, whereupon the pair annihilates itself in the creation of two more gamma rays.

Another hit by a proton makes light helium, He-3, and yet more energy. A collision between two He-3 atoms finally creates normal helium, He-4, with a couple protons left over. The Sun cannot blow up like a hydrogen bomb because the first reaction is creepingly slow. The energy gradually works its way out through successive absorptions and re-emissions by intervening atoms as the gamma rays are gradually turned into sprays of yellow-white light to be emitted at the lower-temperature convective surface. All this activity does not mean that gravity is unimportant. To the contrary, it's gravity's job to provide the compressive force needed to get the temperature and density high enough to run the fusion chain, which in turn provides support and keeps the Sun and similar stars from collapsing. There is enough hydrogen left in the core to keep the Sun going pretty much as it is now for another five billion years. As we will see, the devil is in the "pretty much."

Even among atomic particles, neutrinos, which represent the nuclear "weak force" (as opposed to the binding "strong force"), are strange. The Sun is so dense inside that energy transfer by photons (in which gamma rays are broken down into the more benign visual radiation that warms us) can take hundreds of thousands of years. The sunlight you see today was created all that long ago. But the neutrinos zip right out. Everything is transparent to them (as vividly expressed in John Updike's "Cosmic Gall"). Billions going near the speed of light pass through you every second, day and night, right through the Earth. But we've learned to catch a few with neutrino telescopes, which are as weird as the particles they try to detect. The first was a hundred-thousand gallon vat of chlorine-based cleaning fluid buried a mile deep in South Dakota to protect it from unwanted background radiation. A tiny number of chlorine atoms were changed by neutrino collisions into radioactive argon atoms, which could be counted. Pure water can be used as well. From the rates of collision and capture between neutrinos and various atoms, we find that the numbers coming out of the Sun are just those expected, verifying theory. We can actually peer deep into the solar core and even take images of it. Nuclear fusion works.

Outside the Sun

Back now to the opaque surface, the photosphere, where the solar radiation is released to us. Even a quick look at the Sun usually shows dark patches, sunspots, which range from motes not much bigger than the granules to monsters much larger than Earth. From their spectra, we find that they encompass intense magnetic fields thousands of times stronger than the one generated by our planet. Looking rather like fried eggs, the spots have depressed dark central umbras with temperatures about 1500 Kelvin cooler than the surrounding photosphere. Since brightness is wildly dependent on temperature, the spots appear dark. Surrounding the umbras are lighter, grayish, penumbral slopes, which rise upward to the bright surface. Sunspots appear in pairs with opposing magnetic fields, one

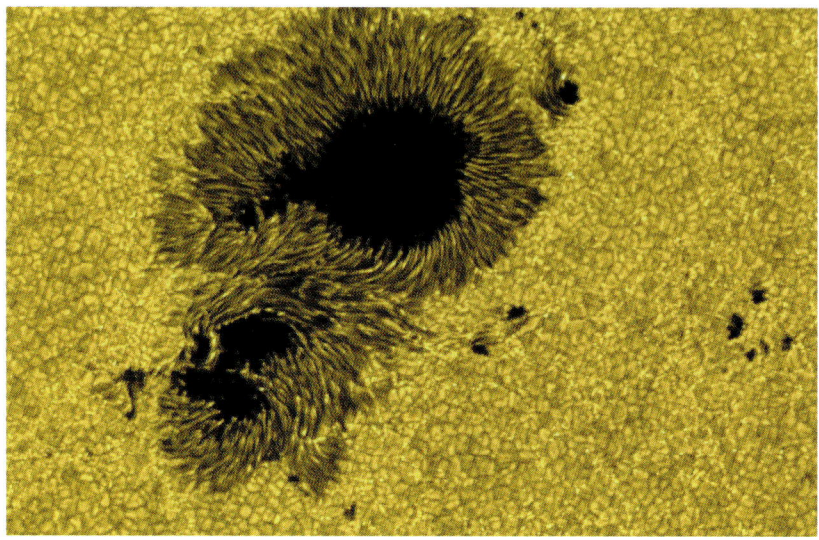

Figure 1.6. The solar surface is made of the tops of millions of convection cells within which hot gases rise then fall. Though appearing small, each cell is the size of a large US state. The huge sunspots to the left anchor vast magnetic loops that extend outward from the Sun and also block the convection. NASA/JAXA, Hinode spacecraft.

positive, the other negative, the pairs commonly assembled into huge confusing groups, into centers of activity that are hard if not impossible to disentangle.

From the rates at which the spots seem to march across the solar surface, we find the Sun's rotation period to be 25 days near its equator, but closer to 30 days nearer the poles, the solar gases madly tearing themselves past one another. The movement of electrically-conducting charged gases through rotation and convection produces a global magnetic field that is concentrated into thick subsurface ropes by the shearing rotation. Floating upward, the ropes pop through the surface in giant loops. Where they enter and exit, the magnetism is so powerful that it inhibits the upward convection, resulting in a pair of dark spots.

In ways still not well understood, the loops and magnetic energy create and heat a vast surrounding envelope, the solar corona, to around two million Kelvin. It's so hot that atoms are ionized multiple times, with many electrons stripped away. Emission lines of highly ionized iron and nickel were once thought to be coming from a previously undiscovered chemical element that was named "coronium," which we know now does not exist. The visible part of the corona seen during a total solar eclipse is about twice the solar diameter, but it extends vastly farther, in a sense even to Earth. People, including scientists, travel thousands of miles to see and study it. The corona is so vacuous, however, that it does not behave like a blackbody and, in spite of its high temperature, is invisible against the blue sky. Close examination from the ground and from space shows it to be filled with hot, bright, looping structures caused by the same magnetic fields that produce the sunspots. Sandwiched in between the photosphere and the corona is a thin, red, magnetically active layer, the chromosphere, the "sphere of color."

The giant loops retain the corona and inhibit it from expanding, as would befit a hot gas. Where the magnetically heated corona is not sufficiently confined, it blows outward in a solar wind. Blowing at a rate of about 10^{-14} of the Sun's mass per year, the wind might seem weak, but it's enough to have a profound impact on Earth. As the wind tears past us the Earth's magnetic field helps create the two surrounding donut-shaped van Allen radiation belts (after James van Allen of the University of Iowa, 1914–2006) of trapped particles, one 1.5 Earth radii out, the other hovering at 4. They in turn protect us from the invading and dangerous solar wind. Funneled down the poles of our planet's magnetic field, the wind's particles make the upper air glow as the northern and southern lights (aurorae) that are readily visible throughout Alaska and northern Canada.

The loops are terribly unstable. If two of them connect and neutralize each other, they violently collapse, sending atomic particles smashing downward onto the chromosphere to create a bright spot, a solar flare, while other particles move outward at high speed to hit Earth. Released, a freed blob of coronal gas can take a couple days to get to us, whereupon it disrupts the Earth's magnetic field, producing aurorae that can stretch to lower latitudes far south of their normal homes. The electromagnetic effects on the Earth's upper

Figure 1.7. An X-ray image of the Sun shows the hot (two million Kelvin) thin gases of the solar corona that are confined by magnetic loops. At the base of each loop is a sunspot. At lower left a loop has collapsed, launching a blob of coronal gas outward. If the Earth were to get hit, a northern lights display would be visible even at temperate latitudes. In an electronically driven world, coronal mass ejections can cause considerable damage. SOHO/ESA/NASA.

ionized atmosphere play havoc with long-distance radio communication, and on the ground are strong enough to bring down power grids, which can cause millions of dollars in damage. The blast can also destroy satellite electronics, though those (and astronauts) that orbit inside the van Allen Belts are at least somewhat protected. Our technological world is frighteningly subject to space weather.

The magnetically driven action goes in cycles that average 11 years in length. At peak activity, the Sun can be quite covered with spots, while at minimum they can entirely disappear. Aurorae, coronal mass ejections, flares, and a variety of other magnetic phenomena follow suit, the whole thing controlled by the time it takes for the deep magnetic ropes to reach their greatest complexity, break down, and dissolve. When they come back, all the fields are reversed, yielding a magnetic cycle of 22 years.

All this impacting energy is somehow absorbed by the Earth, perhaps influencing its climate. Sunspots, which are just symptoms of the underlying solar magnetism, have been linked to everything from drought cycles to the stock market. What seems to be a real effect is related to the "Maunder minimum," when between 1650 and 1715 the spots and cycle disappeared and the northern hemisphere of Earth was plunged into the "Little Ice Age" (which, some argue, could also have been the result of global volcanic activity; just because two things happen at the same time does not mean one causes the other). Other stars show such cycles, from which we can get some kind of prediction. We are maybe due for another shutdown. Now wouldn't that mess up global warming!

2

The Wanderers

From our parochial perspective, the most important feature of the Sun is its set of planets, one of which is ours, the Earth. All shine by reflected sunlight. Traditionally, since the discovery of the outermost one in 1930, there have been nine to memorize, to learn about, Mercury through Pluto. However since the discovery of many more Pluto-like bodies in the far reaches of the Solar System, Pluto's been booted out of the pantheon, so the formal number has been cut to eight. But be careful saying "Pluto's not a planet anymore," as children will assume that somebody blew it up. Pluto's still there! It's now seen as belonging more to a vast zone of orbiting debris, the "Kuiper Belt" (after Gerard Kuiper, 1905–1973), outside the orbit of Neptune, which has become the last planet standing. In a very real sense, the planets are the satellites, the "moons," of the Sun, all born together four and a half billion years ago. While they play little role in the lives of the stars themselves (including that of the Sun), the planets are of great importance in illuminating the way in which the stars were (and are still being) formed. Moreover, the planets are obviously crucial to *us*, to humanity, as one of them provides a home for life. Perhaps the most powerful driving force in astronomy involves the answer to "Are we alone?" Are there others like us? To know, we first have to explore other stars to see if they too have planets, more significantly planets that might also support life (Chapter 8).

Equally important, the Sun's planets are the vehicles that drove the study of gravity, one of the four forces of nature (the others the electromagnetic, weak, and strong forces of Chapter 1). The quest has taken centuries and continues yet today. The planets led to Newton's discovery of how gravity behaves and played a major role in the development of the ultimate theory of gravity, Einstein's relativity. Of the four forces of nature, the first to be "discovered" (by the first human to fall off a cliff), on an atom-to-atom basis, gravity is by far the weakest and perhaps the most mysterious, as no particle is known to be associated with it (as photons are with the electromagnetic and others are with the weak and strong). So in any discussion of the Sun and other stars, we can hardly ignore the planets, and instead must embrace them as central to our story.

More true facts

Within the Solar System, distances are measured not so much in kilometers, but in Astronomical Units, the AU, the average distance between us and the Sun, 149.6 million kilometers, or 93.0 million miles. The size of the Solar System depends on how you define it. The planets spread out to 30 AU, the distance from the Sun to Neptune. But that's hardly the end of things. The Kuiper Belt (which contains Pluto) goes out to at least 45 AU, while various rogues climb to hundreds of AU away. The census is not yet in, but there are several bodies that rival Pluto in size (including some that go well beyond the Kuiper Belt) plus millions of smaller ones, their number depending on the lower cutoff in radius (do you count a dust grain?). The influence of the solar wind, to where it butts up against the interstellar gases, takes us more than thrice as far as Neptune, to 100 AU, a region explored by the ancient (launched in 1976) Voyager spacecraft that long ago imaged the giant outer planets. Never to return, they will orbit our galaxy of stars forever, making an announcement (one likely never to be heard) that we exist. Farthest afield, trillions of icy potential comets orbit the Sun in loneliness within the "Oort Cloud" (after the Dutch astronomer Jan Oort, 1900–1992), which seems to extend to 100,000 AU or more, a significant way to the nearest star. It's given away by the comets that leak from it into the inner Solar System.

Sizes and distances are difficult to embrace. A scale model helps. Make the AU 100 meters long, a distance roughly of that standard unit of modern measure, the "football field," whatever kind is played. (The Greeks beat us to it, their "stadium" the length of a ... stadium.) The Sun is now a ball 0.9 meters wide. Few would notice us, as the Earth becomes a rock just 0.85 cm (a third of an inch) across. A celestial balance would put the Sun in one pan, 330,000 such rocks in the other. Here is home to all our lives, hopes, and dreams. Mercury, a stone less than half the Earth's size, is 39 meters from the solar goalpost, while Venus, just a bit smaller than Earth, is 72 meters out. Masses scale comparably. Another small stone just a bit bigger than Mercury, Mars, is in the parking lot 150 or so meters away from the Sun.

We then take a giant jump to Jupiter, half a kilometer (nearly a third of a mile) away. The largest of the Sun's family, Jupiter is a ball of gas and liquid hydrogen (don't pick it up) some 10 times the size of Earth, a tenth that of the Sun (in our model 10 cm across), and is more than 300 times heavier than our planet. After passing Saturn, almost a kilometer (0.6 mile) away and roughly comparable to Jupiter, we run into a much smaller pair (4 Earth radii, 15 Earth masses), Uranus and Neptune. Vistas now become vaster, Uranus 1.9 kilometers (1.2 miles) away from the scaled Sun, Neptune more than half again farther. Between Mars and Jupiter is a sand and gravel pit of debris (the asteroid belt). Another beyond Neptune includes Pluto, which is but 1.5 millimeters (just over an eighth of an inch) across. It's amazing that it was found, and nearly a century ago at that. On this scale, the Oort Cloud of potential comets starts to approach the size of our actual Earth.

Moving right along, the nearest star, Alpha Centauri, is 28,000 kilometers (17,000 miles) away. That's twice the diameter of Earth, nearly 30 million times the solar diameter, and a typical interstellar distance. While it does not look that way on a dark, star-filled night, space is as empty of stars as atoms are of "matter," of protons, neutrons, and electrons. It's no wonder that stars don't collide. And that space travel is so daunting. Humans have been to the Moon, just a quarter of a solar diameter away. To get to the nearest star we have more than a 100 million times farther to go. Even if the Voyagers were pointed in the right direction it would take tens of thousands of years for them to cover the distance. It's unlikely that we are going anywhere soon. Or for that matter that anyone has ever come to us (to be continued…).

The planets are not all lined up like they are on a school poster, but go around the Sun on more or less circular orbits, counterclockwise as viewed looking downward from the north, and in similar planes with low tilts (inclinations) relative to Earth's path, all of which provide profound clues as to planetary, even solar, origins. The farther they are from the Sun, the slower they move and the farther they have to go. The orbital periods, the times they take to go once around, thus depend critically on the distance from the Sun, and range from just 88 days for Mercury to 165 years for Neptune, and (why not?) 239 years for Pluto, so they will typically be found all over the place. As seen from Earth, Venus will be in one direction, Mercury in another, Jupiter in yet another. As they orbit, they quite obviously move against the background stars, hence their collective name from the Greek "planetai" for "wanderers."

The actual numbers are given in the table below, which for each planet gives the average distance from the Sun in AU, the diameter and mass relative to Earth, the orbital period in days or years, the viewing cycle (the "synodic period," to be visited later), the tilt of the planetary orbit relative to Earth's and the orbital eccentricity (the departure from circularity). The Earth for example changes its distance from the Sun by just 1.7 percent in each direction. The most circular orbits are those of Venus and Neptune, which change their distances by just a percent or so. Note too that the tilts are low, meaning the planetary orbits closely define a plane, which is critical to what follows. The big exceptions to the low tilts and eccentricities are Mercury and Pluto, both of which augured revolutions in thought, Pluto to what it means to be a "planet," Mercury to Einstein's relativity. Mars is a bit off too. But it's pretty clear that Pluto doesn't fit at all.

Table 2.1. Planet facts.

Planet	Distance (AU)	Diameter (Earth's)	Mass (Earth's)	Orbital period	Viewing cycle[a]	Inclination (degrees)[b]	Eccentricity
Mercury	0.39	0.38	0.055	88 d	399 d	7.0	0.21
Venus	0.72	0.81	0.81	225 d	584 d	3.4	0.01
Earth	1.00	1.00	1.00	365 d	—	0.00	0.017
Mars	1.53	0.53	0.11	1.88 y	2.13 y	1.9	0.09
Jupiter	5.2	11.2	318	11.9 y	399 d	1.3	0.05
Saturn	9.6	9.4	95	29.4 y	378 d	2.5	0.06
Uranus	19.2	4.0	14.5	84.0 y	370 d	0.8	0.05
Neptune	30.1	3.9	17.1	164 y	367 d	1.8	0.01
Pluto	39.4	0.18	0.002	249 y	367 d	17.1	0.249

[a]As seen from Earth.
[b]Relative to Earth's orbital plane.

As seen from Earth

Imagine the (almost) spherical Earth in space with you on top, the way it appears to you. Those on the other side are not desperately hanging on. Drawn by gravity toward the Earth's center, everyone feels on top. For the sake of argument, you stand somewhere in the mid-northern hemisphere. Now put the sky, the infinite "celestial sphere," around it. Expand a plane at your feet tangent to the Earth's curve and it slices the sky in half at the perfect astronomical horizon. Ignore mountains, trees, houses, and cows. Everything above the horizon is visible, everything below is out of sight, as the Earth gets in the way. Directly overhead lies your zenith. Since we live on a spherical planet, everyone has his or her own personal zenith. As you move, it goes along with you. Position on Earth is defined by latitude and longitude. Latitude is the angle north or south of the equator, longitude the angle east or west of the prime meridian, a north-south line through an observatory in Greenwich, England, as defined by international agreement. The Earth rotates daily about an axis that exits at the north and south poles, latitudes 90 degrees north and south. Run the rotation axis of Earth outward, and it strikes the celestial sphere at the north and south celestial poles (NCP and SCP). Above and parallel to the Earth's equator (zero degrees latitude) runs the celestial equator, the sky simply reflecting the Earth. The celestial equator intersects the horizon at due east and west, defining the two compass points, which are actually on the sky, not on Earth. (You can never get to a place called "East": it keeps moving in front of you.) Running perpendicular to the equator through the celestial poles and your zenith is the celestial meridian, which hits the horizon at due north and south, so defining the other two compass points, giving us all four. The NCP, by accident closely marked by the star Polaris (within a degree), is elevated above your horizon by an angle equal to your latitude, the first rule of celestial navigation. At the north pole of Earth, 90 degrees north latitude, the NCP is at the zenith, 90 degrees up, while at the steaming equator, latitude zero, the celestial poles lie on the horizon and the celestial equator passes overhead through your zenith. At 45 degrees north latitude, look north and halfway up the sky to see Polaris.

While the Earth rotates counterclockwise, from west to east, it does not feel that way, so eschew reality for what seems to be. In response to the Earth's rotation, the sky appears to turn about us in the other direction, clockwise (no coincidence), carrying the stars, Moon, Sun, and planets, with it. Watch as they rise over the eastern half of the horizon and set past the western half. If a star is sufficiently close to the north celestial pole, it is always visible, neither rising nor setting (circumpolar), while if it is close enough to the south celestial pole it is forever out of sight unless you travel and change the celestial sphere's north-south orientation.

The Earth orbits the Sun once a year, every 365.2422... days. There is no link between daily rotation and annual revolution; they are independent of each other. Pretend however that we are stationary, and that the Sun seems to go about *us*, which will also be in the counterclockwise direction. The Sun will appear to trace a path, the "ecliptic," to the east against the distant stars. If you walk around a chair in the middle of the room, it seems that the chair is going around you against the family pictures.

If the Earth's axis were bolt upright to the plane of its orbit, the ecliptic would lie directly on the celestial equator. The Sun would then rise due east, set due west, and always pass the same angle above the horizon at noon. But it isn't and it doesn't. Instead, the rotation axis is tilted relative to the orbital perpendicular by an angle of 23.4 degrees, another "accident of nature" that probably goes back to the origin of the Earth and Moon (and all the rest of the Solar System) some four and a half billion years ago. The ecliptic must then also be tilted by the same angle relative to the celestial equator. The Sun follows this tipped path to the east against the stars at a rate of about a degree per day (the close match between the number of degrees in a circle and the length of the year is no accident), while at the same time going about the sky on a daily path caused by terrestrial rotation. The Sun crosses the celestial equator (when it is directly overhead at the Earth's equator) about March 20 at the Vernal Equinox and September 23 at the Autumnal Equinox, which respectively mark the beginnings of northern hemisphere spring and fall. As the Earth rotates, on these dates the Sun indeed rises due east and, after tracing out the celestial equator, sets due west (leaving out some technical details like the atmospheric refraction of Chapter 1). Since the Sun is on the equator, and the equator is split evenly by the horizon, the Sun is up for 12 hours, down for 12; days and nights are thus of equal length, hence "equinox."

The tilt takes the Sun from the Vernal Equinox up to 23.4 degrees north of the equator on June 21 to the Summer Solstice, when from temperate northern latitudes it is highest at noon, then down past the Autumnal Equinox to 23.4 degrees south of the equator on December 21 at the Winter Solstice, when it is lowest. Then it's back to the Vernal Equinox to start the cycle all over again. Here is the sole cause of the seasons, as in northern summer the solar light impacts the Earth from on high with maximum warming efficiency, while in winter it strikes at a shallower angle, spreading itself out, resulting in a low heating rate. The Earth's southern hemisphere sees the opposite.

The Sun and ecliptic pass through a set of 12 ancient constellations collectively called the Zodiac. The Vernal Equinox is in the

direction of Pisces (the celestial fishes), the Summer Solstice in
Gemini (the Twins), the Autumnal Equinox in Virgo (the Maiden),
the Winter Solstice in Sagittarius (the Archer). Since the orbits of
the planets are all close to the same plane, the planets will usually be
found within the Zodiac as well. The Moon, which has a 5.5 degree
orbital tilt to the ecliptic, goes through them as well. Indeed, there
are 12 constellations of the Zodiac because to the nearest whole
number there are 12 phase cycles in the year. At each full Moon, or
other specific phase, the Sun is one constellation over to the east.
Since the planets orbit counterclockwise as seen from above the
north pole, they generally move from west to east against the stellar
background. The motions of nearby Mars and Venus can be seen
night-to-night. Outer ones go more slowly, Jupiter taking 12 years to
go around at a rate of one constellation per year, while Saturn's
journey is nearly twice as long.

But since we view them from a moving platform, our Earth, their
apparent motions are more complex. As Earth orbits, we first catch up
with an outer planet, pass it when it appears opposite the Sun, then
pull away from it, only to catch up with it again in a perpetual viewing
cycle, or synodic period, which is just the lapping time. It's 2.1 years
for Mars (the fastest of the outer planets), then approaches a year as
we go farther out to the slower ones. The two planets in the inner solar
system do the reverse, lapping us, coming (though rarely exactly)
between us and the Sun. When the Earth goes between an outer
planet, or a speedier inner one passes us, it appears to go backwards,
to the west against the stars, or "retrograde." Given the ancients' belief
that the Sun and planets all went around a central Earth, these movements were a source of consternation, and the early astronomers
constructed elaborate schemes to explain them. Then came reason.

The revolution and the gang of six

It's only natural to think that the Earth is in a central position, the
center of the Universe. Much of astronomy over the past few centuries has been devoted to displacing Earth (and ourselves) from one
"center" to another until none is left and we see that there is no

center at all. There are many who contributed to understanding how we think it all works. But six stand out.

The first of the bunch was the Polish cleric, Nicolaus Copernicus. Born in 1473, 19 years before Columbus "discovered" America, he spent much of his life demonstrating that the Earth and planets could all go not about Earth, but about the Sun. His great work, "de Revolutionibus Orbium Celestium" (in which he gave the distances of the known planets from the Sun in AU), was published in 1543, the year of his death, which may have saved him from the Inquisition.

Next up is Tycho Brahe. Born in 1546, just after Copernicus died, Tycho was perhaps the greatest naked-eye observer who ever lived. The Danish king set him up with an island observatory that Tycho called Uraniborg. With giant protractors, he measured the positions of hundreds of stars and recorded the movements of the planets against them. Position in the sky is in modern terms defined by "declination," the angle north or south of the celestial equator, and "right ascension," the angle east or west of the Vernal Equinox. You could write a book on it. (Your author DID write a book on it.) A contentious man, after a change in the regime he forcibly left for Prague in 1597. Dying in 1601, the results of his studies were turned over to (purloined by?) his assistant, our number three man Johannes Kepler (1571–1630), a German mathematician who had gone to Prague to work with Tycho and who turned the tide. Copernicus's heliocentric theory did not predict the motions of the planets any better than had those from the older geocentric theories. Like his predecessors, he had made the logical error of having the planets move in circular orbits, the sphere and circle being the perfect figures, ones that would naturally have been chosen by God.

The ancient Greek belief is that we could know the Universe through pure thought. In the spirit of modern science, Kepler did the opposite and set himself to calculate how the planets actually move so as to replicate the data, specifically Tycho's observations. It took nearly a decade before he would announce his first two laws of planetary motion, which were based on the rapid movement of Mars, and yet another decade to pronounce the third law, which included the other planets.

Kepler's first law of planetary motion is that a planet moves not in a circle, but along an ellipse with the Sun at one focus. No one before had ever considered such a deviation from presumed perfection. The ellipse is drawn so that the sum of the distances along the curve to two interior points on the long axis, the foci, is a constant. Bring the foci together and you have a circle; stretch them way apart relative to the size of the figure, and the ellipse comes out long and skinny, almost a straight line. The distance between a planet and the Sun must therefore continuously change. The Earth is nearest the Sun (1.7 percent closer than average) at perihelion, about January 2, farthest at aphelion, around July 4. In the northern hemisphere, we are closest to the "fire" in the dead of northern winter, farthest in the heat of summer: so much for that theory of the seasons. See above. All things being equal, summers and winters in the southern hemisphere should be somewhat more intense than those in the north, but the effect is pretty much quashed by the distribution of the heat-storing oceans, which dominate the south.

Equally important, in his second law Kepler found that planets do not move uniformly along their elliptical paths. Instead, they speed up as they approach perihelion, slow down as they recede to aphelion. Properly stated, the "radius vector," the line that connects the planet to the Sun, sweeps out equal areas in equal times, so as the distance to the Sun goes down, the velocity goes up and vice versa. The two laws together explain the "inequality of the seasons" that puzzled the ancient Greeks. Though it may not seem so (as those who hate the cold will tell you), northern winter, when the Earth rushes through perihelion, is three days shorter than summer, when it drags through aphelion. Similarly, fall is shorter than spring.

Finally, Kepler connected all the planets with number three, his Harmonic Law: the square of the orbital period of a planet (P) in years equals the cube of its semimajor axis (a) in Astronomical Units (the semimajor axis rather obviously being half of the ellipse's long axis). $P^2 = a^3$. Try it. Jupiter is 5.2 AU from the Sun. Cube it ($5.2 \times 5.2 \times 5.2$) to get 140.6, the square root of which is 11.9, the period in years. "You can look it up." Kepler rests his case.

The fourth person in this starry hit parade is Galileo (1564–1642), who began his quest to know the heavens in 1609, about the time Kepler was presenting his first two laws. Galileo did not invent the telescope, but with the best ones he could make turned to the sky found: (1) the moons of Jupiter, and that they went around the planet much as Copernicus's planets go about the Sun; (2) craters and mountains on the Moon; (3) sunspots, the latter two showing that neither the Moon nor Sun is perfect; (4) the phases of Venus, which are not possible as seen in the old Earth-centered system without a lot of reconfiguring; (5) that the Milky Way is made of lots of stars. There's much more including the rings of Saturn, which he mis-interpreted with his small scope as knobs.

What really set him apart from others who were doing similar work is that he continued with his observations, sought to interpret them, realized that he had given proof to Copernicanism, and wrote extensively and positively about it. Which inevitably got him in trouble with the Church. Called to Rome in 1633, he was forced to recant and was placed under house arrest for the remainder of his life, which ended in 1642. His works were banned until the nineteenth century.

But "the cat was out of the bag" (an old naval phrase having nothing to do with actual cats). And nobody could put it back in. Keplerian theory worked beautifully. But at the same time nobody could figure out WHY it worked. Kepler thought some mysterious force pushed the planets around. It remained for numbers 5 and 6 in our tale to get it right.

The last two told how it all works

It's said that Isaac Newton was among the most brilliant people who ever lived. Contentious, reclusive, with few friends, his mind ranged over a vast tract of intellectual space. He, in parallel with Leibniz in Germany, invented calculus, which he developed from the concepts of Descartes (from whom arises another ancient joke best left untold). He also invented the reflecting telescope, was among the first to examine the solar spectrum (Chapter 1), and in one of his greatest achievements (all described in his "Principia") set down laws of motion and gravity, and in doing so also explained the tides.

An acceleration is any change in velocity, which as a technical term combines both speed and direction. Increase or decrease your speed or drive around a curve in the road at constant speed and you are accelerating. Newton postulated three laws of natural motion.

(1) A body's motion stays constant unless acted upon by an outside force. A ball will sit on a table forever unless you whack it. But then to the contrary, the ball slows and stops. Friction, an outside force, is the culprit. Something always gets in the way of the ideal, which is what makes mechanics so interesting.

(2) The acceleration achieved by a body is directly proportional to the force applied to it and inversely proportional to the body's mass (thus defining the meaning of "mass"). Push a football player and a jockey with the same force. Guess which will go farther, faster. Then, running in the other direction, apply the first law to yourself. If acceleration is force over mass, force = mass times acceleration, or as it is usually stated, $F = MA$.

(3) Every action has an equal and opposite reaction, which causes airplanes and rockets to work. The hot gas goes out the end of a jet engine, and it and the attached fuselage go in the other. With Galileo as predecessor (story is that he dropped balls of different weights from the Leaning Tower and found they landed at the same times, and if it's not true, it should be), Isaac Newton invented physics.

No apple hit Newton on the head, nor did he "discover" gravity. Gravity is a staple to all things on Earth, including ourselves. He did something far greater by discovering how gravity works. Newton noted that the acceleration of a falling body (which is independent of mass, see Galileo) and the acceleration of the orbiting Moon were in inverse proportion to the squares of their distances from the center of the Earth. What makes the apple fall is a universal force that also makes the Moon follow its closed curving path.

Ignoring any outside forces (quite impossible actually to do), the force of gravity between spheres is directly proportional to the product of their masses and inversely proportional to the squares of the distances between their centers. Or $F = G(M_1 \times M_2)/D^2$, where "$G$," found in the lab, is a constant that makes the units come out right. Terrestrial gravity behaves as if the Earth's mass were all concentrated to the center of the planet. The part of the Earth closer to you exerts more force, while that farther away exerts less, so it all comes out even. In the sense of the law, you are 6500 km (4000 miles) from Earth, our planet's radius. As you stand, the Earth pulls you down toward the center, but since the Earth is solid you can't get there. Your "weight" is thus a force against the surface of Earth such that $W = G\, M(\text{Earth}) \times M(\text{you})/(\text{Earth radius squared})$. Substitute the mass and radius of a planet for those of Earth, and get your weight on Mars or the Moon. (What would you weigh on Jupiter? Nothing. There is no solid surface and you'd fall into it.) Since $F = MA$, the acceleration of gravity (the acceleration with which you or anything else falls) is G times the mass of Earth divided by radius squared, or $g = G\, M(\text{Earth})/R^2$. It's measured at 9.8 meters per second per second (32 feet/s/s). Jump off of something. The longer the duration over which you accelerate, the faster you will go, your speed increasing by 9.8 m/s for every second you are in the air. Better to jump from a chair rather than the top of the building. Skydivers hit a terminal velocity of a couple hundred miles an hour, compliments of air resistance and Newton's first law. Measure "g" and the radius of the Earth (first done by Eratosthenes in the third century BC from changes of the position of the Sun as one moves north or south) and you can find our planet's mass.

What about the tides mentioned above? The lunar gravity exerted on the part of the Earth directly under the Moon is higher than it is on the Earth's opposite side. The result is a stretching across the Earth, noticeable mostly in its oceans, which will have a bulge that in principle faces the Moon. As the Earth turns, a point on the shore will go under deeper then shallower water a bit more than twice a day. In reality, high and low ocean tides lag behind the position of the Moon in the sky. The Sun also gets into the act with a tide of its own about a third the size of that produced by the Moon. At new and full Moon they add together, while at the quarters the solar partially fills in the lunar.

Two bodies placed in proximity to each other will always raise mutual tides. Tides raised in the solid body of the Moon by the Earth have stopped its rotation relative to us such that we always see the same lunar face. If you were on the Moon, the Sun would rise and set every couple of weeks, but the Earth would hang nearly unmoving in the lunar sky. At the same time, tides raised on Earth (including solid-body) by the Moon are slowing our rotation by about a thousandth of a second per century, which messes a lot with keeping precise time, leading to "leap seconds" and another book.

But, as they say on TV commercials, "Wait, there's more!" Edmund Halley, he of comet fame, showed that Newton's theory predicts elliptical orbits. Newton said that he'd already shown that. What exactly then is an "orbit?" Drop a ball and time its fall to the ground. There is no relation between horizontal and vertical motion. Throw the ball horizontally and it immediately starts to fall at the same rate at which you first dropped it. The faster you throw it though, the farther it will go before landing. At some speed, the curvature of the Earth comes into play. Throw the ball at 18,000 miles per hour and it drops at the same rate at which the Earth curves below. Though falling constantly, it can't catch up with Earth's surface and just keeps going around and around (presuming no friction with air and that neither mountains nor Grandma gets in the way). The ball's now in orbit. At the Earth's surface, the circuit would in principle take 88 minutes. Put an astronaut into orbit a hundred miles up, where it really works, and she and her spacecraft fall with the same acceleration. She feels "weightless" even though she's within the Earth's gravity, which has not dropped off by very much. (If you still have good knees, jump again from the chair. While you are falling, you weigh nothing.) Following Kepler's third law, orbital period squared is proportional to semimajor axis cubed, so the farther the satellite is from Earth (pursuing a circular path), the longer it takes to orbit. At a distance (from Earth's center) of 41,000 km (25,000 miles), the period is the same as the

Earth's rotational period of 24 hours. If you place the thing above the Earth's equator, it then seems to hang motionless in the sky and we have found the best place to put a communications satellite.

Using his laws of motion and the law of gravity, Newton came up with generalizations of Kepler's Laws.

(1) The orbit of one body around another is a "conic section" with the first body at the focus. Cut a cone at different angles. Perpendicular to the axis you get a circle; tilt it some and you get an ellipse. Tilt the cut parallel to the side, and out comes a parabola, while a greater tilt yields a hyperbola. Parabolic and hyperbolic orbits are open-ended, one way: here comes the deadly asteroid; there it goes, never to return.

(2) The conservation of angular momentum. Tie a rock to a string and swing it around your head. The angular momentum is the rock's mass times its speed times the string's length. Change one quantity and you automatically change another so as to maintain constant angular momentum; pull in the string and the rock goes faster. As a planet gets closer to perihelion, it speeds up in response to the shortening distance, while as it approaches aphelion, it slows down, satisfying Kepler's original second law. Watch a dancer or skater behave the same way as she brings in her arms.

(3) Gravity involves mass, yet mass does not appear in Kepler's third law, $P^2 = a^3$. Kepler used units relative to Earth, periods in years and distances in Astronomical Units. Newton not only saw how the planetary masses fit in, but expressed the third law in physical units. He found that the squares of the periods, P^2 ("P" now in seconds), equals a constant times a^3 ("a" in centimeters, meters, or whatever) divided by the sum of the masses of the two bodies, or $P^2 = (\text{constant} \times a^3)/(M_1 + M_2)$. (The constant is $4\pi^2/G$, where π ("pi") is the ratio of the circumference of a circle to its diameter, 3.14159...) If M_1 is the mass of the Sun and M_2 the mass of a planet, $M_1 + M_2$ is essentially constant as well. Kepler got away with his third law because planetary masses are so small compared with the solar mass: the mass of

the Sun plus that of the Earth does not change much if the Earth is replaced by Jupiter. It's a powerful rule. Tip it upside down and you can use the orbit of the Earth to measure the mass of the Sun or the orbit of one of Jupiter's moons to find the planet's mass, etc.

But there is also a powerful caveat. This treatment works only if the mass of the orbiting body is inconsequential. What if it is not? In reality, one body never actually orbits another. To say it does is a mathematical simplification. Orbits are always mutual, each body going around a common center of mass between them whose position depends on the inverse of the mass ratio ($a_1/a_2 = M_2/M_1$, where a_1 and a_2 are the semimajor axes of masses 1 and 2 about the center of mass). As Moon orbits Earth, Earth orbits Moon, though with a smaller orbital radius of just 3000 miles as determined by the mass ratio. Say the individual masses are comparable. The fiction that the smaller body goes exclusively about the bigger (with a semimajor axis of $a_1 + a_2$) gives the sum of the masses. The location of the center of mass (which can be difficult to find) then gives the mass ratio, hence the individual masses.

Number 6? Albert Einstein (1879–1975), who showed that Newton was wrong, though not by enough to prevent the greatest triumph of Newtonianism.

Expanding the solar system

In 1781, William Herschel (of infrared fame; see Chapter 1) nearly doubled the size of the Solar System. While systematically scanning the skies with one of his telescopes, he found what at first he thought might be a comet. But regular observations by him and others clearly showed its planetary nature, a body about a third the size of Jupiter at 19 AU out from the Sun in an 84-year orbit. The discovery caused a sensation. Herschel originally named it "Georgius Sidus," "George's Star," after his monarch, King George III of American Revolution fame. While George was pleased, the French and others thought it was not such a good idea, so all eventually agreed to continue the mythological sequence, Uranus (the god of the sky) being the father of Saturn.

The Sun clearly has major control over its planets. But each planet must influence all the others, both directly and indirectly through its gravitational interaction with the Sun. Though the tug of one planet on another is small, it's relentless. That makes for a lot of interconnections when one is to calculate real planetary orbits, which then cannot actually be perfect ellipses. There is no general mathematical formula to give the mutual orbits of three bodies, let alone many. To calculate an orbit, first locate all the planets, then figure the mutual accelerations on all of them to see where they are going to be tomorrow, then re-calculate the accelerations, and so on ad infinitum. It's extremely difficult to do without a computer, but nineteenth century astronomers became masters at it.

Uranus is not all that faint, and under good conditions can be seen with the naked eye. Calculations of its orbital path showed it had been observed as early as 1690 by others and thought to be merely a faint star, so there was a string of data to work with that included post-discovery observations. And Uranus was never just where it was supposed to be. But if there is a planet beyond Saturn, might there be one past Uranus? John Adams in England and Urbain Leverrier in France set out to "look" for it, each trying to predict the position of the presumed trans-Uranian planet on the basis of the deviations it would produce in Uranus's calculated positions, which critically included the effects of the known planets. In a long, complicated, and disputed story Adams's work was not followed up. Leverrier sent his prediction to the Berlin Observatory, where Johannes Galle found what was to be named Neptune right away, on September 23, 1846. Thirty AU from the Sun, it has just completed one full 165-year orbit since it was found.

The discovery caused yet another sensation. Newtonian physics works! If you knew the positions and velocities of all the atoms in the Universe, you could predict the future. Determinism is king. "Your honor, I robbed the bank because Newton's Laws made me do it." The philosophy had some credit until the advent of quantum mechanics in the early 1900s and the discovery of the natures of subatomic particles. Like photons, electrons and protons behave with wave–particle duality such that we have no hope of pinning a "particle" (such as it is) down. All we can give is probabilities as to "position." Free will, or perhaps probability, therefore now reigns.

But even factoring in Neptune's pull, Uranus did not perfectly behave itself, so the same trick was tried again. Finally, in 1930 Clyde Tombaugh of the Lowell Observatory in Arizona came across Pluto. Terribly faint, the "last planet" (so-called at the time) turned out to be far too small to influence much of anything other than its own moons. The discovery had been fortuitous. Averaging 40 Astronomical Units from the Sun, taking 249 years to go around, Pluto has a highly tilted, eccentric orbit that for 20 years takes it inside that of Neptune. Moreover, Neptune has a lock on it, as Pluto — about the size of the western US — orbits twice for every time Neptune orbits thrice (a word you don't get to use very often). The bigger planet has a lock on the smaller. Pluto thus turns out to be the main member of the Kuiper Belt (the ring of planetary debris outside the orbit of Neptune), and was thus "de-planeted" to a lower status. Or perhaps raised to a higher one. Improved masses of the outer planets from the movements of passing and orbiting spacecraft finally solved Uranus's misbehavior. There seem to be no more actual planets, no "Planet X." Though hold that thought as the "rogues" mentioned above still suggest the possibility of ultradistant planets, perhaps one as big as Earth.

Returning to the inner Solar System and the nineteenth century, Newtonian theory had another problem. The pulls of all the planets cause the orbit of Mercury to rotate slowly counterclockwise. Newtonian calculation predicts that the Mercury's perihelion should go around the Sun at a rate of 530 seconds of arc per century. Frustratingly, the path of the gods' messenger persisted in rotating 43 seconds of arc faster. Perhaps yet another planet inside Mercury's orbit was to blame. It was even given a name, "Vulcan." But where was it? Even during solar eclipses it could not be found. Again, no "Planet X." Something was wrong. The solution lay elsewhere, with the last of the gang of six.

Albert

More disturbing than Mercury's problem was that of the speed of light. It was once thought that light waves were transmitted through a vague, all-pervasive substance called the "aether" (which nobody had ever actually detected). In 1887, Albert Michelson and Edward Morley measured the speed of light first along the direction in which the Earth was orbiting, and then perpendicular to it. If the aether existed, the two measures should differ by the Earth's orbital velocity of 30 kilometers (18 miles) per second. They didn't. Not only is there no Planet X, there's no aether either (though at least 70 years later it was still being taught in schools).

The explanation for both problems lay in Einstein's theory of relativity, which holds that we live in a web of four-dimensional spacetime (a combination of three-dimensional space coupled with one of time) in which the speed of light is constant no matter what the observer's velocity. If you want to hop a freight car being hauled from Chicago to New Orleans (not recommended), you first run faster and faster until you match its speed and climb aboard. You can't do that with light. No matter how fast you run, it just keeps passing you at the "speed of light." To an outside observer, your mass also increases, so it requires progressively more energy to move ever faster. To actually hit light speed, at which point your perceived mass would become infinite, would require infinite energy, which is not available. Only massless photons can get there, hence "c" is the speed limit to the Universe. To maintain the constancy of the speed of light, the runner's time must slow down. At the speed of light itself, time stops. The name "relativity" comes from the realization that the timing of an event is always relative to the position and motion of the observer. We all see the Universe slightly differently and, because of the finite speed of light, at different times.

Einstein posited a "principle of equivalence," that mechanical acceleration and the acceleration of gravity act the same way. Nobody knows why. The classic "thought experiment" uses an elevator deep in interstellar space away from all gravitational influences. (Odder things have happened.) Step inside, the doors close, and an

invisible hand draws the chamber upward with a constant acceleration of 9.8 meters/second/second, the acceleration of gravity. You
feel like you are back home in Earth's gravity field. You cannot tell
the difference. A narrow beam of light suddenly pierces the darkness. Since the elevator is accelerating, the beam appears to bend
downward, which then must also happen in a gravity field. But a
light beam always defines a straight line (the shortest distance
between two points). What you feel as gravity is interpreted as the
bending of spacetime. In lower dimensions it's rather like putting a
bowling ball on bed spring: everything rolls toward it. Gravity has
energy. But $E = Mc^2$, so gravitational energy has a mass equivalent,
which changes the rules a bit, though there is far more to it than
this. When Mercury's orbit is calculated with the new theory (now a
century old), we recover the missing 43 seconds of arc. The effect is
not so noticeable in the other planetary orbits because they are farther from the Sun and move slower.

There is hardly a scientist alive who would not like to disprove
the theory. In spite of countless tests, nobody's done it yet. Relativity
is essential to modern life. Without it, we would have no successful
planetary probes and the Global Positioning System (which relies on
a large set of Earth-orbiting satellites) would not work. Nor would
the calculation of planetary orbits.

The planets near the Sun

Planets are a natural part of the star-forming process. To even begin
to understand those orbiting other stars, which are legion, we need
first to understand something of our own. We've looked at the planetary system and how it works. We finish quickly by examining the
individuals in much the same way a Reduced Shakespeare Company
does all the plays in a few minutes.

The first four planets, Mercury through Mars, bear superficial
resemblance to one another. Called the "terrestrial planets" after the
largest of their membership, Earth, all are rocky with nickel–iron
cores. Earth's core takes up about half the planetary diameter. As
found from earthquake vibrations, the core's inner portion is solid,

the outer liquid. The outer half of Earth, the mantle, is mostly silicate rock. On top floats the solidified rocky crust, which is divided into lighter-weight continents and low water-filled basins. Though the oceans look huge, they are no more than a thin film a couple of miles deep on top of a body 8000 miles wide. We are really quite dry, though nowhere nearly as much as the other terrestrials, including Mars, whose water is seemingly constantly in the news.

The Earth's interior is heated to thousands of degrees through its ancient formation and by the radioactive decay of heavy uranium and thorium into lead. Circulation of the liquid core produces our magnetic field and ultimately the aurorae. The mantle is in a hot plastic state that slowly flows in convection currents that break through the crust mostly at sub-oceanic ridges. Fracturing the crust into individual plates, the currents push the continents around, making them crash into one another to create mountains, earthquakes, and volcanos. Volcanos like those that made the Hawaiian chain also arise directly from the mantle's depth well away from plate boundaries. Above it all lays our oxygenated 80-percent nitrogen atmosphere.

The closest body to us is the Moon (see Figure 1.4). Circling Earth, it's not considered a "planet," though it surely looks like one. About a quarter the size of Earth, with a bit over one percent of Earth's mass, it's a beaten, battered body covered with impact craters and lunar "maria" (Latin for "seas"), which are large basins coated with ancient dark solidified lava that make the face of the "Man in the Moon." Most of the craters were dug out some four billion years ago by collisions with the debris that followed on the heels of Solar System formation. The basins, which are mostly huge craters, came along shortly thereafter through a series of gigantic strikes. They are oddly on the side of the Moon facing Earth (the Moon not rotating with respect to us, compliments of tides). Since then, the cratering rate has dropped substantially, as witnessed by the relatively few craters within the maria.

If the Moon is so cratered, how did Earth escape? It didn't. The heavy cratering was wiped away by erosion and continental drift. The 200 or so known impact craters are relatively recent. Some of the impacts were so powerful as to have wiped out most of the terrestrial species alive at the time, including the one 65 million years ago that destroyed the dinosaurs (probably by producing so much dust that it altered the climate). Too small to hold on to any significant insulating atmosphere, lunar surface temperatures range from above the boiling point of water down to 150 below Celsius, or even lower. There is little water. The Moon is almost completely dry with just a bit of ice within deep polar craters that see no sunlight. With a tiny frozen iron core, the Moon has no global magnetic field, which allows the solar wind and coronal mass ejections to pound down on the surface, making it a terribly dangerous place to live.

A day on the Moon[1]

It seems likely that one day we will return to the Moon to set up permanent bases. Lacking air, our satellite is an impressive observatory site, and the farside is shielded from interfering radio radiation from Earth. What would a lunar day be like?

Pick a spot — for example, Mare Vaporium, near the center of the lunar disk as seen from Earth. The Sun would first appear when the Moon is in what we see as first quarter. It would be night even minutes before sunrise since there is no air to scatter sunlight and create twilight. The landscape would still be relatively bright, however, because of the reflected light from the third-quarter Earth, which would hang stationary in the lunar sky. Your first hint of impending sunrise would be a glow from the solar corona, seen on Earth only during a solar eclipse. Then the tip of a mountain or crater wall to the west would suddenly light up as it caught the first direct solar rays. Since the lunar day is 29.5 Earth days long, the Sun would creep up at 1/29 the pace it does on Earth; at the Earth's equator it takes the Sun two minutes to vault the horizon; on the Moon it would take about an hour.

Very slowly the temperature of the surface rocks would climb, an increase that would be of little consequence to any residents, since most would be living underground to shield them from the effects of both temperature and the solar wind. It would actually be rather dangerous to spend too much time on the surface because of this high-speed particle radiation. Only a meter or so down, however, the Moon's temperature remains quite stable and no particle radiation can penetrate.

A week later, it is finally noon and the "new" Earth has disappeared. Another week, and the Sun begins to set. It turns dark as soon as the solar disk disappears, except for the now-brightening Earth and a few peaks that catch the last of the solar rays. Now you settle down for two weeks of long, cold lunar night. This pattern would be repeated day after lunar day, as it has for billions of years.

Figure 2.1. Like the Moon, Mercury is plastered with impact craters that go back to the early days of the Solar System. While too small to retain an atmosphere, permanently-shaded craters near the poles protect deposits of ice. Mercury Messenger, NASA/JHAPL/Carnegie Institution of Washington.

Mercury, with a diameter just 38 percent that of Earth, is the smallest planet (barring Pluto and its kind). At first it appears similar to the Moon. Close to the Sun and hot, surface rocks hit 430°C (near 800°F) at noon, but they drop to under 180 below (−300°F) in the dead of night. It too has little of any atmosphere, just that formed by atoms kicked off the surface by the solar wind. It's so close to the Sun that it's hard to see from Earth, rising and setting only in twilight. It's even too angularly close to the Sun for the Hubble Space Telescope to look at. What we know has come from passing and orbiting spacecraft. Like the Moon, Mercury is covered with impact craters (and basins) that go back to the early days of the Solar System, though it also has extensive inter-crater volcanic plains and odd sinkholes. Mercury's biggest physical feature is its huge iron core, which takes up some 70 percent of the planet's radius. The original mantle may have been ripped off in an ancient collision, which would also explain some of the planet's odd surface chemistry. Small amounts of ice again reside in dark polar craters. The iron core should by now have frozen solid, yet Mercury possesses a weak magnetic field that is offset from the center. Like Moon to Earth, Mercury is tidally locked onto the Sun but, because of the rather high orbital eccentricity, with a rotation period (relative to the stars) of 59 Earth days, exactly two-thirds of the 88-day Mercurian year.

A day on Mercury[1]

It is highly unlikely that anyone will set foot on Mercury in the foreseeable future. Living conditions would be formidable, not only because of the searing heat of the Sun, but also because of the solar wind, which is energetic enough to create the bulk of the sparse Mercurian atmosphere. Yet the idea of living there provides an interesting thought experiment. The Sun would average 1 1/4° across, 2.5 times greater than its angular diameter seen from Earth. The Sun's angular size would also noticeably change as Mercury passed from perihelion to aphelion, from 1.5° to 1°. The solar day, 176 Earth days long, would proceed with agonizing slowness. At that pace, the Sun would move along its daily path at an average rate of only 2° per Earth day, or a mere five minutes of arc per Earth hour. Watching a sunset or sunrise would be an exercise in extreme boredom, the whole event lasting on the average 18 Earth hours. Stranger still, the apparent daily solar motion would depend strongly on Mercurian longitude. As Mercury approaches perihelion and begins to move faster, the Sun's east-to-west daily motion would slow, and for a short time, the Sun would actually appear to go backward. Think now of living at a point where the Sun rises or sets near perihelion. There is a specific longitude where at sunrise, the solar disk would make an appearance, reverse direction and set in the east, and then come back up again! Other than that bizarre event, the day would proceed much as on the Moon, but with surface temperatures climbing to the lead-melting point. After an equally long sunset, the nighttime temperature would plunge to −185°C. Of course, you could live underground where the temperature could be maintained fairly comfortably with a little air conditioning, but then you would see nothing at all. It would not be a pleasant place to live, or even to visit.

Often called Earth's twin, Venus is just a bit smaller. Some twin. It rotates backwards with a period of 243 days relative to the stars, 117 days relative to the Sun, probably as a result of coupling with the thick carbon dioxide atmosphere, which blankets the planet with a pressure 100 times that of Earth's. The greenhouse effect has gone mad, boosting the surface temperature to close to 470°C (near 900°F). Covered by thick sulfuric acid clouds, the rocky surface is invisible to the visual observer but has been closely examined by orbiting craft using radar. The old Soviet Union even landed probes on it. The planet displays myriad volcanoes similar to mid-ocean Hawaii and a crater count that suggests the planetary surface volcanically repaved itself no more than a billion years ago, which wiped out any early record. Extreme volcanism may have cooled the core enough that the planet produces no magnetic field. With no

Figure 2.2. Venus is covered with opaque clouds swirled by slow rotation and made of sulfuric acid and other noxious chemicals that float in a carbon-dioxide atmosphere 100 times the pressure of Earth's. Mariner 10/NASA/C. J. Hamilton.

real continents, with lower basins unfilled with absent water, the place is an astronaut's nightmare.

A day on Venus[1]

With a pressure suit and an intelligent choice of location — perhaps a pole, where the sunlight slants obliquely — Mercury might be at least livable for an astronaut. There is even water available from the thin ice caps. One really wonders about Venus. It seems unlikely that human beings will ever walk the planet. The conditions — the heat and atmospheric pressure — are too severe. Nevertheless, radar images and Venera (Soviet lander) pictures show a fascinating landscape. What would it be like to stand in Lakshmi Planum and gaze at Maxwell Montes off in the distance, or to climb Rhea Mons in Beta Regio to look into the rift valley, or even to try to walk across the tesserae (landforms)? You would certainly have a long enough period of daylight, 58 Earth days, to make any journey you wished. Unlike days on the Moon or Mercury, however, your day would start with a long twilight because of slow rotation and sunlight scattering from the high, thick atmosphere. Of course no stars would ever be visible at night, and come daylight all you would see would be a high, probably featureless, cloud deck. To your initial confusion, the sky would begin to brighten in the west because of Venus's retrograde spin. It would never get very light outdoors and would probably appear somewhat as it does at home under a very heavy thunderstorm cloud deck, the kind that brews tornados. The sunlight that got through would be heavily absorbed and would have an orange cast to it, giving the landscape an eerie burnished glow. As the long day proceeds, you would have to be careful never to venture out without a protective suit. Not only could you not breathe, but you would literally be cooked in the searing heat, comparable to the temperature of an oven in a self-cleaning cycle. You would not have much weather to worry about, however; the forecast would be the same day after day, hot and dry, with a slight wind from the east. For all the evidence of volcanic and meteoric activity on the planet, you would never likely experience any. The craters have had half a billion years to accumulate, and geologic processes are slower than they are on Earth. As the Sun sets in the east, it would gradually get darker but it would never cool off. The air is too good an insulator. You would probably want to leave as quickly as you could to go home to the wet, blue, cool Earth.

Smaller Mars, about half our size and ten percent our mass, is far more hospitable. Its rotation period and axial tilt (24 hours 37 minutes, 25.2 degrees), and thus seasons, are similar to ours. At the poles are thick ice caps that during fall and winter advance toward lower latitudes. Some astronomers once believed they saw fine lines, "canals," crossing the surface. Canals, of course, could only be built by Martians. Alas, both are illusions. The southern hemisphere is populated not by people but by ancient craters, while the northern is known for volcanoes, one of which towers more than twice the height of Everest. You can build bigger mountains on lower-gravity planets, especially if there is no continental drift to limit them. Mars tried to be geologically active, with a rift 3000 miles long, but cooled off too quickly inside for the planet to make continents. The air, with just under one percent the pressure on Earth, is almost all carbon dioxide. The polar caps are mostly water ice with some dry ice in the south. The low atmospheric pressure cannot support significant liquid water, but there is plenty of evidence that it once was prevalent. We see branching dry valleys in the south, huge eroded channels in the north, possible shorelines, suspiciously layered rock, and evidence for an aqueous chemistry. The place must once have had a thicker atmosphere and been much like Earth. Was there once life? There is no evidence of any. And if not, why not?

Figure 2.3.   Mars has a landscape sculpted by wind and by water that once flowed freely and is now trapped underground and as polar ice. There is no evidence for life. But since the planet was at one time something like ours, why not? Curiosity, NASA/JPL-Caltech, MSSS.

A day on Mars[1]

Your spacecraft takes you to a location near the equator of the red planet during the southern-hemisphere summer. A day there would certainly be more pleasant than one on Venus. You could not go outside without protective garments, however: the air is too thin to support human life, and there is no oxygen to breathe. Furthermore, the Sun shines almost unimpeded by the atmosphere, so lethal ultraviolet rays reach the ground. Nevertheless, you could get out of your spacecraft and easily walk around. Other than the need for a spacesuit, a Martian day might seem rather like one spent in a high, very dry, cold desert like central Antarctica. At sunrise, preceded by a weak twilight, you see frost or maybe a little fog. After sunrise, these watery effects would rapidly dissipate. You might extract some ice from underground to bring back to the ship for drinking water. It is a cloudless day, though airborne dust tints the air a bit pink. The wind blows gently from the south, and by noon the ground temperature has risen nearly to the freezing point of water. As you lightly walk — you weigh 37 percent what you do on Earth — the regolith crunches under your feet, and you occasionally kick up little puffs of dust. The wind rises and you see a few dust devils, small whirlwinds, playing off in the distance. The Sun proceeds on its daily path, as at home, and near sunset you return to the ship to turn up the electric heat and prepare for a temperature drop of 100°C. While you anticipate the next day of exploration, Deimos, the smaller tiny outer moon, shines faintly, like a slowly moving star. Perhaps Phobos, the larger nearer moon, both probably captured asteroids, will rise in the west, making you long for the more familiar Moon of home.

Here there be giants

Traveling farther outward, toward the bigger planets, we first encounter a host of very small ones, the "minor planets," the asteroids. Hundreds of thousands, millions, are spread mostly between 2.1 and 3.2 AU from the Sun. The largest, Ceres, is just under 1000 kilometers (575 miles) across, well below 10 percent the size of Earth. Collisions and planetary gravity throw a few as far out as the orbit of Jupiter and farther inward than Earth. Small ones that hit us are called meteorites. Big ones are trouble. Meteorites come in two basic flavors, rocks and irons, telling us that the asteroids must be similar. Some of the rocks are primitive, going back to the birth of the Solar System 4.5 billion years ago. Indeed, it's the radioactive dating of primitive meteorites that gives us the age of the Solar System. The irons must be the fractured remains of asteroids that, like the Earth, differentiated into iron cores surrounded by rocky mantles. Many are surprisingly icy. Over the aeons big asteroids occasionally and disastrously strike us. We are trying to track the dangerous ones down with the dream of altering the orbit of any predicted to hit Earth.

Then come the lumbering giants, Jupiter, at 5.2 AU, the largest, 11 times the size of Earth, carrying more than 300 Earth masses. Like the Sun made largely of hydrogen (though in the molecular form of paired hydrogen atoms) and helium, the planet is covered with ammonia clouds full of noxious hydrocarbons that stripe out parallel to the equator because of strong shearing winds. Surprisingly, Jupiter spins with a "day" less than half Earth's. Most likely there is nothing big enough to slow it down. Storm systems abound, led by the seemingly permanent Great Red Spot. More than twice the size of Earth, a high-pressure cyclone rotating with a period of six days, it's been around for over 300 years. Deep down, the compressed hydrogen turns liquid, the circulation of which generates a magnetic field more than a dozen times the strength of Earth's. Filled with trapped solar particles, the Jovian magnetosphere is not only lethal, it's among the brightest sources of extraterrestrial radio emission seen from Earth. It also produces powerful polar aurorae.

Four large satellites orbit the planet. Discovered by Galileo, entertaining to watch as they go around on the order of days, the satellites can be seen in binoculars. Innermost is Io. Slightly bigger than the Moon, with an orbital period of only 1.8 days, it's just six Jovian radii out. Gravitationally tugged about by the next two,

Figure 2.4. Jupiter's colorful clouds, made mostly of ammonia crystals mixed in with various hydrocarbons, float in a hydrogen–helium atmosphere. The Great Red Spot at lower right, an anti-cyclone, is at least 300 years old and could easily swallow Earth. Carrying 300 times Earth's mass, Jupiter would barely notice. Hubble Space Telescope, NASA/ESA and A. Simon, Goddard SFC.

Europa and Ganymede, tides raised by Jupiter flex and heat its interior to make Io the most volcanically active body known as it spits out plumes of sulfur-laden silicates. It's so close to Jupiter that there is an electric current running between the two. Only a little smaller than Io, Europa is quieter, though tidally heated to the point where it seems to embrace a warm ocean below a surface made mostly of ice. Could it be an abode for life? Icy Ganymede, the size of Mercury, is the largest satellite in the Solar System, while Callisto (26 Jovian radii out, taking about a fortnight to orbit) is not far behind. A common theme out here is ice, which makes up about half the mass of the outer big satellites. The space around Jupiter is shared by more than 60 tiny moons, some probably captured asteroids.

A visit to Jupiter[1]

A trip to Jupiter sounds interesting, but would you really want to go? The conditions make the worst of the terrestrial planets seem benign. The journey alone is arduous. Like the Voyager craft, you might cruise across interplanetary space for almost two years. That is survivable; after all, the crews of tiny sailing vessels did much the same regularly in the seventeenth and eighteenth centuries. Arriving at Jupiter, where would you land? There is no accessible solid surface. Progressing inward, you would encounter a slushy beginning of a molecular hydrogen ocean, but no person or machine could actually survive even part of the passage. There is a spot among the clouds where you could drift happily in a balloon at a comfortable temperature and pressure, although attempting to breathe the hydrogen/helium/hydrocarbon atmosphere would quickly prove fatal. Instead, you go on to the satellites. Io would be a fascinating stop. You land on the side facing the giant planet, which appears 20° across in the sky, larger than the Big Dipper seen from Earth. You can rather easily watch the huge planet rotate. You orbit it in less than two days, the stars rising and setting much as they do on Earth, while immense Jupiter hovers stationary in your sky because of tidal locking. In the distance you may see a geyser filled with sulfur dioxide spewing straight up, like the stream from a giant firehose, so high it disappears. You may even overlook the banks of a molten sulfurous river or lake. To appreciate the eerie beauty of the place, however, you would have to be swathed in heavy shielding to protect you from energetic particles. It is doubtful, in fact, that you could take shielding heavy enough. Proceed to the other satellites and Jupiter retreats into the distance. From bright, shiny Europa it has shrunk to 12° across, and from grooved Ganymede to 8°. Even there, the particle radiation field would probably be deadly. On ancient Callisto, the magnetosphere weakens to a point where you might survive, but for long-term safety you would need to withdraw to one of the tiny outer satellites, where the planet you want to study is about the size of the full Moon in our own heavens. Better to stay home and send a robot.

Figure 2.5. If there is an astronomical icon, it's Saturn, a planet almost as big as Jupiter but with a third of its mass. Saturn's graceful rings, nearly 300,000 kilometers (175,000 miles) across, are made of countless rocky iceballs and seem to be the result of a tidally-disrupted moon. The big gap, the Cassini Division, is caused by the repetitive gravitational pulls of one of Saturn's larger satellites. The outer narrow "Encke gap" derives from an embedded satellite. All four of the giant planets have ring systems made of collisional debris. Fainter rings extend many times farther out. Only a few hundred meters thick, the inner ring system is the thinnest-known thing in the Universe. Hubble Space Telescope: NASA and E. Karkoschka (U. of Arizona).

Then near 10 AU from the Sun we find glorious Saturn, which, like Jupiter, is made largely of hydrogen and helium. Though carrying just a third the Jovian mass, the lower gravity allows it to puff up to nearly Jupiter's size, making it much less dense. Even a cursory look reveals a similar banded structure of ammonia clouds, though more obscured by methane haze. While the magnetic field is much weaker than Jupiter's, it's strong enough to produce aurorae. Storm systems come and go.

But mostly, Saturn is known for its system of flat, ultrathin but bright rings whose diameter is double that of the planet. Made of particles of ice and rock typically a few centimeters in diameter, the rings are embedded with larger bodies that, along with the gravitational action of other satellites, drive a great and beautiful structure of nested ringlets. The easily-visible rings, probably made of a smashed or tidally disrupted satellite, are just the inner part of a vastly extended system. Orbiting outside the obvious rings are some

20 modest satellites and far more smaller ones. Many have bizarre characteristics, like Iapetus with two distinct faces and Enceladus with watery volcanic plumes. The largest moon, just under Ganymede's size, is strange Titan. With a thick, visually-impenetrable atmosphere, it's covered by clouds that rain liquid methane into hydrocarbon streams and lakes, as actually seen by the probing Cassini spacecraft.

Saturn and the joy of science[1]

Like little else, examination of Saturn by the Voyager mission experts demonstrates that science is not a dry, dispassionate study but an exciting exploration into the unknown. Try to imagine what it must have been like for the people at the video consoles as Voyager 1 approached its destination. The words of Bradford Smith, the Voyager imaging team leader, express it best: "As the final days of the approach phase were upon us, anticipation grew. The satellites were finally being seen as more than just individual disks and there was something very peculiar about the bright rings — they were daily showing more and more structure, far more than could be accounted for by simple satellite resonance theory. We were preparing for an exciting encounter, but even bigger surprises were to come." Later he notes, "Mimas seemed to be more normal — except for a giant impact crater that is more than a third the diameter of Mimas itself. As the image of Mimas and this absurd crater appeared on our television monitors, there was a sense of déjà vu. Of course! It was George Lucas's 'Death Star!'" And still later: "... our attention was fixed on the F ring. Resolution was improving rapidly, and the apparent decreasing width of the ring kept pace... those of us watching the monitors were stunned. If ... we had become somewhat jaded to the unpredictability of the outer Solar System, our sense of astonishment was brought back in an instant... Staring back at us from the television monitors were three individual strands, each approximately 20 km wide and separated by a few tens of km; they appeared to be knotted, kinked, and braided. To me, it was the most improbable picture yet sent back by either Voyager spacecraft."[1]

Figure 2.6. Neptune, four times the size of Earth with 15 times the mass, is denser than Jupiter and Saturn. Its methane haze features the Great Dark Spot, which has disappeared since the image was made by Voyager 2 in 1986. Bright cirrus clouds float at its periphery. Highly tilted Uranus is similar but with fewer features. NASA, JPL.

We might take Uranus and Neptune, at 19 and 30 AU, together, though each has highly distinctive characteristics. Both are much less massive, smaller, and denser than lightweight Jupiter and Saturn and, though filled with hydrogen and helium, contain much more water/methane/ammonia "ice." Both have highly tilted magnetic fields offset from the planetary centers that are significantly weaker than Earth's. Uranus's chief oddity is that its rotation axis is tipped over almost perpendicular to its orbital plane, giving it extreme seasons. Next is probably a set of narrow, dark, widely spaced rings almost the reverse of Saturn's. Neptune and Jupiter have thin ring systems as well. All are debris belts of some sort. While Uranus has a large set of smallish satellites, Neptune has few, though one large moon, Triton, stands out. Orbiting backwards, it's almost certainly a captured body, one that looks a lot like Pluto.

Launching outward

So what's Pluto? Along with its large satellite and four dinky ones, Pluto is not only orbitally-locked to Neptune, its apparent one-time brother, Triton, is now actually owned by the planet. Pluto was the "last planet" until late in the twentieth century when other smaller bodies, some in orbits similar to Pluto's, began to turn up. Dozens were found, then thousands. One, Eris, has properties that rival Pluto's, as do a few others, including Triton. If Pluto is a planet, why not Eris, why not the myriads of much lesser bodies? Pluto, now considered a "dwarf planet" (as is the largest asteroid, Ceres), is the harbinger of a collection of small, icy rockballs that extends out to about 55 AU, with a few scattering well beyond. The existence of these objects was predicted by the Irish astronomer Kenneth Edgeworth (1880–1972) and later by Gerard Kuiper as the reservoir of short-period comets, those that take less than 200 years to make a full journey around the Sun.

Figure 2.7. Like the biggest asteroid, Ceres, Pluto, has been removed from planetary status, in part because of many similar bodies. The "dwarf planet" is only about the size of the western US. The first images by New Horizons revealed an unexpectedly complex surface with ice flows and mountains as well as both smooth and cratered terrains. New Horizons, NASA/JHUAPL/SwRI.

Comets are icy, dusty bodies on long elliptical orbits. As a comet approaches the Sun, it partially "melts," the ice subliming directly to gas, which is then ionized by sunlight. The solar wind subsequently blows the gas backward into a long, bluish, glowing tail. The melting of the icy matrix releases the dust and small rocks. Sunlight both reflects off the particles and pushes them backward into a yellow-white curved dust tail. Comets do not streak across the sky but, like planets, move in stately order a bit each night, usually getting brighter as they approach the Sun, fading as they recede, though there are wild exceptions. As augured above, there are two broad kinds of comet, short and long period. The short kind stick to the ecliptic plane and orbit counterclockwise like planets. Frequently encountering the Sun, they evaporate quickly and are generally faint.

Collisions and the gravitational effects of the outer planets gradually move Kuiper Belt objects inward until they visit closely with the Sun and become actual comets. There must be billions of them out there. That said, Pluto, Eris, and the other bigger ones, are not comets, but are evolved bodies that perhaps "tried" to become planets but just could not gather enough material to grow to decent size through constant collisions and mergers with smaller bodies.

About once a generation, we are visited by a truly great comet that has an orbit that takes thousands, even millions, of years to complete. Encountering the Sun infrequently, perhaps even for the first time, a long-period comet can be amazingly bright; some have even been visible in daytime. Among their company is Halley's. Its orbit was worked out by Edmund Halley, who found that different historical comets were actually the same one coming back over and over again. With a rather short 76-year period, it's an exception to the short-period comets, its original orbit perhaps clipped by a planetary encounter. Halley's lore is deep. It appears on the Bayeux Tapestry and has a dreadful reputation as a bringer of evil. The last time it came by, in the mid-1980s, Mexico City fell down in a great earthquake. It's been accused of carrying off small children. To the contrary, comets, great and small, have no influence on Earth, as their gravity is far too small. Unless they, like asteroids, hit us, as

Comet Shoemaker-Levy did Jupiter in 1994. A lesser comet caught by the giant planet, it was pulled apart into multiple pieces through Jupiter's tidal action, then one after another they struck. If there, why not here? Many of the impact craters on Earth should have been caused by comets. Unlike asteroids, the appearances of the long-period comets are quite unpredictable, giving us little time to save the planet. It's widely believed that early collisions with comets (and watery asteroids) brought the cooling Earth its water. Even if they don't hit anything, comets that periodically revisit the Sun are doomed to death by evaporation.

The long periods and huge orbits told the Dutch astronomer Jan Oort (1900–1992) that there must be another comet reservoir filled with a trillion icy bodies that could extend outward perhaps

Figure 2.8. A great comet like Hale–Bopp comes to us from the vast Oort comet cloud about once a generation. As it gets close to the Sun, its binding ices evaporate to produce a blue ionized gas tail that is swept downstream by the solar wind. Released dust and small rocks form the yellowish dust tail. Short-period comets, most quite small, are stored in the Kuiper Belt beyond Neptune. Cometary debris (as well as asteroidal junk) hitting the Earth's atmosphere are seen as meteors that streak across the sky. J. B. Kaler.

halfway to the nearest star. Encounters with stars and interstellar clouds occasionally bring a few inward to appear as the long-period comets. Back in the Kuiper Belt, Eris goes out to 100 AU, another, Sedna, out to nearly 1000. Are we beginning to see Oort Cloud comets, or things related to them, in place? And what else might be out there? From these distant bodies, we make a great leap to the stars, to other planets, and perhaps other life.

Life at the end of the Solar System[1]

We earlier spent a day on blistering Mercury, which is tucked in so close to the Sun that lead would melt on its surface. What is the other extreme like? Travel now to an outpost on Pluto, near the end of our planetary system, where we stare out into the depths of interstellar space. The trip alone is daunting. It took Voyager 2, with accelerating gravitational assists by Jupiter, Saturn, and Uranus, a dozen years to make the journey to Neptune, while it took New Horizons nine years to make it more directly to Pluto. Finally, you arrive on this dim, cold world at a time when it is at its average distance from the Sun, 40 AU. Put on your pressure suit and climb out of your craft. A strange Sun shines in a black sky. There is no apparent disk! To your eye, it appears only as a blinding star. Look away and a huge, dim quarter moon (Charon) more than 3 degrees across hangs over the horizon. Crunching across a nitrogen–methane snow frozen to only a few tens of degrees above absolute zero, you arrive at your cabin, where you can finally breathe oxygen without your space suit.

Your day is a long one. Very slowly, the Sun moves across the sky. It was noon when you arrived; 1.5 Earth days later, the distant star has finally set. Although the looming moon steadily goes through its phases, it stays motionless in the sky, the captive of mutual synchronous rotation. Night sets in and the temperature drops even further. At least your view of the stars is good.

Time to call Earth. Pick up the phone and start talking. No one answers. You talk for about an hour and a half, telling the mission commanders or perhaps your family what it is like 6 billion kilometers from home. You hang up. Your "Hello, I'm fine" has just reached the orbit of Neptune. It will arrive at Earth nearly five hours after you began speaking. Six or seven hours after that, your phone rings.

Gaze outward into space, away from the Sun. There are no planets going through their familiar oppositions. Look toward the Sun. All planets are inferior (closer to the Sun). With luck you might glimpse Jupiter at greatest elongation a mere 7 degrees from the daylight star. Earth is almost lost. In your loneliness, you pull out your telescope for a glimpse of home. There it is, a fragile "pale blue dot" (Sagan) against the frigid sky. It's going to be a long year.

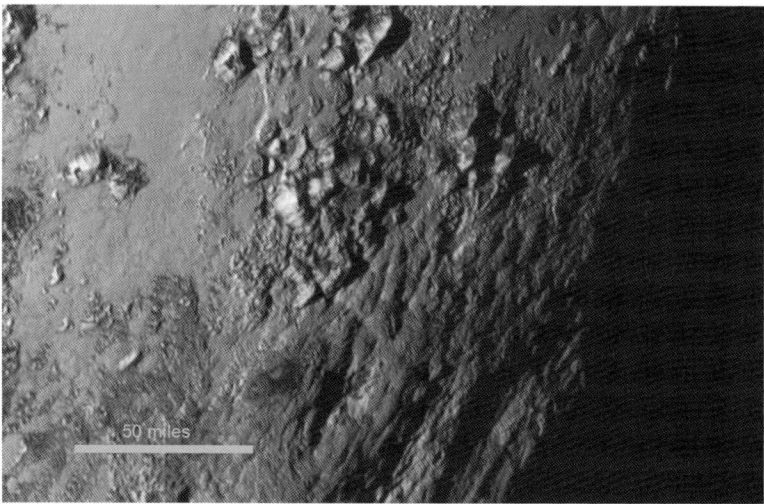

Figure 2.9.   Among Pluto's many features are Rocky-Mountain-sized peaks made of ice. Orbiting within the Kuiper Belt, which is the source of short-period comets, Pluto is more evolved. We are not close to inventorying the Solar System, let alone the vast spaces that lie in the great beyond. New Horizons, NASA/JHUAPL/SwRI.

Reference

1. J. B. Kaler, *Astronomy!*, HarperCollins, New York, 1994.

3

Seeing Far

"You can observe a lot by watching." (Yogi Berra, who said he did not say all the things he said.) And watch we do, for thousands of years, both to enjoy and to comprehend the sky, its stars, its planets, its galaxies. Hardly trivial, pre-telescopic observation provided the foundation for Copernicanism, Kepler's Laws, and ultimately for part of Newton's great work on the rules of physics. But there are stringent limits as to what the unaided eye can accomplish. The breakthrough was made in 1609, when Galileo turned his telescope onto the sky. Though it had a lens only a couple inches across, it provided a powerful new technology that allowed him to reap vast rewards, which included evidence that Copernicus was not just dreaming, but correct.

What telescopes actually do

First, they make stars, planets, etc. brighter by gathering more light than the eye can possibly do alone. Simply put, a telescope is a light bucket that collects starlight with lenses or mirrors and brings it to a focus for the observer to see directly or to record permanently. We can thus see ever-fainter stars, asteroids, galaxies, whatever, just by making the lens or mirror bigger. Telescopes also magnify, making the scene look larger, as can be attested to by anybody who has used binoculars (twin aligned telescopes) at a racetrack or for that matter anywhere else. To the astronomer, magnification refers to the ability to see or resolve fine detail that the eye alone cannot discern. What astronomical telescopes do NOT do is cut the light-travel time from a star or other object to here.

The past four centuries since Galileo, and especially the last few decades, have seen a stunning increase in our ability to know the heavens, to probe deeply and with increasing detail into the cosmos. Aided by sensitive detectors, powerful computers, and larger telescopes, we can scan the sky nightly to accumulate so much data that we store it for future generations to mine, so extraordinarily much that data management has become an astronomical specialty. Moreover, we have expanded the concept of "telescope" to cover observation of the entire electromagnetic spectrum, from long wave radio to high energy gamma rays, and of necessity have taken our telescopes into space in order to free us from the blurring and absorptive effects of the Earth's atmosphere, which blocks much of the electromagnetic spectrum. So here we look at the traditional tools of astronomy and at how we gather the data through which cosmic understanding (or so we think) might be accomplished. And it all started with Galileo's tiny lens that helped change the world.

Refractors and how telescopes work

There are two different kinds of astronomical telescope used in the optical domain, refractors and reflectors, with many variations on the themes. Refractors use lenses, reflectors rather obviously mirrors. Since the refractor is the more familiar of the two, we'll use it as an example to outline basic properties of telescopes in general. Like Galileo's, the refractor uses a curved, convex lens (one bowed outward in the middle) called the objective. Stars and the like are so far away that their incoming light rays (or photons) run parallel to one another. The lens's purpose is to gather the incoming parallel beams of light and converge them toward a focus where they meet. The distance from the objective to the focal point is the focal length, which depends on the curvature of the objective. If you put a screen at the focus perpendicular to the lens's axis (the focal plane), you'll see a projected image of the source, be it a field of stars, a planet, or perhaps a distant bird, which will appear inverted, up for down, and left for right.

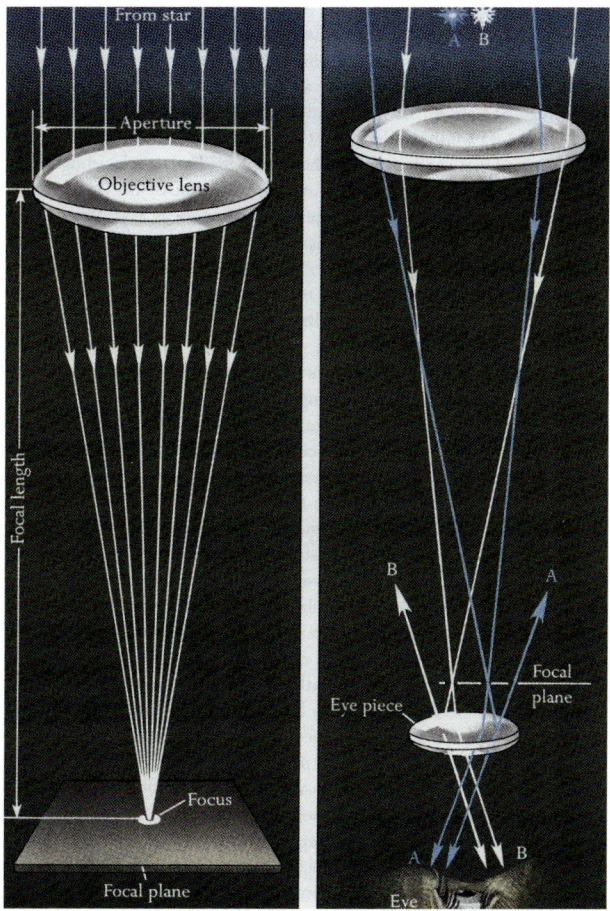

Figure 3.1. At left, a simple lens (the objective) of a refracting telescope brings starlight to a focus. At right two off-center stars are focused onto a focal plane. An eyepiece renders the light waves from each star parallel again for viewing by eye. From "Stars," J. B. Kaler, Scientific American Library, Freeman, New York, 1992.

The image can then be examined with a second convex lens, the eyepiece, which is set in back of the image and acts as a sort of magnifying glass. Before coming to the focus, starlight rays converge, while beyond it they diverge. The purpose of the eyepiece is to render the diverging rays parallel again so that they can be focused by the eye onto the retina. (Galileo used a concave lens, one thinner in the middle, set in front of the focus, which is less effective and no longer done.) The magnifying power of a telescope is the factor by which it makes things in the sky, or on the ground for that matter, look larger, and is the ratio of the focal length of the objective to that of the eyepiece. We can achieve different powers simply by changing eyepieces that have different focal lengths.

That's the way it was done "in the old days" (were they really better?) and still is among legions of amateurs (in the true sense of the word), whether they are observing from the back yard or at organized star parties. The point is to witness the richness of the heavens first hand with one's own eyes and mind. While there is room for some visual observation (of variable stars, of ephemeral events on planets, etc.), nobody actually looks through a research telescope anymore except when you can get away with it to sneak a peek. Instead, some sort of detector is placed at the focal plane so as to permanently record the image to take into the office for measurement and study and to store for later generations.

The most important characteristic of the objective is its diameter, or aperture, which controls the amount of light that can be gathered from the scene and thus the brightness of the image or the ability to see ever fainter stars. The light gathering power (very different from the magnifying power) of an astronomical telescope depends on the surface area of the lens, which in turn depends on the square of the diameter. A six-inch telescope has a lens six inches wide. For a given focal length, a 12-inch lens will make the scene four times ($12/6 = 2$, squared) brighter, which also allows you to see stars four times fainter.

But there is another reason for size, one not often expressed. When passing a barrier, light waves bend and interfere with one another, a process called "diffraction." Even though an objective may be large, its periphery still acts as a barrier. As a result, the image of a pointlike star is actually a small, smeared out "diffraction disk" that is surrounded by faint diffraction rings. The same phenomenon is responsible for the tight rings around the Moon as seen through translucent clouds, or rings about street lights when viewed through fogged eyeglasses. The larger the lens's diameter, the smaller the star's central disk. Larger lenses thus better resolve pointlike sources such as two stars close together and can also reveal more detail on an extended source such as a planet. The minimum separation of two points to be achieved is the "resolving power," which for a 10-centimeter (four-inch) telescope is theoretically 1.5 seconds of arc, but for a one-meter telescope (ten times as large) is ten times as good, 0.15 seconds of arc. Because of aperture's supreme importance, large telescopes are named by their lens (or mirror) diameters, giving us "the 36-inch" of Lick Observatory in California or "the 200-inch" of Palomar.

Refractors have two big problems. Along with refraction goes dispersion. Glass lenses also act like prisms to produce tiny rainbows. The image of a star will as a result actually be a colored line along the lens's axis, with the red image of the star at the farther end, the violet image at the shorter, which makes a mess of the view. If you focus an eyepiece or set a detector on one color, all the other colors will be out of focus and hazy, the result called chromatic aberration. The effect is minimized by employing lenses of low curvature and therefore low refraction and dispersion. Low lens curvature also means very long focal length, hence the huge and unwieldy instruments used by some post-Galilean astronomers. One wonders how such telescopes could even be pointed.

The modern, but still limited, solution is to use pairs of lenses in which a convex lens of one refractive ability is set against a concave one of a different ability such that the whole unit is still convex. You can select different glasses and curvatures so as to bring any two colors to a focus, which makes the lens into an "achromatic doublet." If the telescope is to be used visually, with an eyepiece, typical for the nineteenth century, the red and yellow images are usually brought to a focus. The result is an improved image but one with a bluish, out-of-focus haze around it. The original photographic emulsions, however, were far more sensitive to blue light than is the eye. In the era of photography, achromatic doublets were designed to bring yellow and blue together. Three lenses in a "triplet" can bring three colors to a focus, and so on. But there is a limit, as the more pieces there are, the more the light is absorbed by the compound lens and the dimmer the stars will look, defeating larger aperture. Modern lenses made for amateur astronomers now use non-glassy materials that pretty much do away with chromatic aberration, though cost is high.

The other, perhaps more important, difficulty with refractors is that they can be made only so large. Glass is not entirely rigid. If you (carefully) hold a large piece of window glass at the edges, it will bow in the middle. Large lenses will sag as well, and in different ways as the telescope is moved to point in different directions, which disturbs the sharpness of the image. The largest refractor in the world is at the University of Chicago's Yerkes Observatory in Wisconsin, established in 1897. It has a lens 40 inches (a bit over a meter) in diameter, which is tiny as compared with modern reflectors. Number two, the 36-inch refractor at the Lick Observatory of the University of California (on Mount Hamilton overlooking San Jose), opened to the skies a decade earlier.

The mirrored sky: Reflectors

Both these problems are avoided by using a concave mirror instead of a lens, one similar to a magnifying cosmetic mirror. Light comes in from a star, hits the curved surface, and is brought to a focus in front of the mirror. Since reflection is independent of color, reflectors have no chromatic aberration. To achieve proper focus at the focal point for light reflected from different parts of the mirror's surface, the mirror is usually shaped into a paraboloid, the cross section of which is a parabola, one of the conic sections introduced in the last chapter (odd how quite different things can relate to each other). The flatter the mirror, the longer the focal length. It's an old idea that goes back to before Isaac Newton, who made the first one. Herschel's telescopes were all reflectors. They really began to make their mark, though, in the twentieth century. As to size, "the sky's the limit."

The first telescope mirrors were made of reflective speculum metal, which is an alloy of copper, tin (which together make bronze), and a bit of arsenic. The stuff did not reflect very well, tarnished easily, and required almost continuous care and repolishing, which could easily change the critical curvature. Moreover, arsenic is a bit on the dangerous side. A thin coat of silver on glass is much more reflective. Unlike a bathroom mirror, it's deposited on the front side of the glass rather than on the back, so there is nothing for the light to have to go through. But like silverware at home, it too quickly tarnishes, and has to be frequently replaced. The modern solution is to use a thin film of aluminum, which is deposited on the curved base in a vacuum chamber. Though not as reflective as fresh silver, it oxidizes cleanly and can go for several years before it needs to be redone. Ordinary glass, however, is sensitive to temperature changes, expanding when it heats, contracting when it cools, which is why a cold glass pot can crack when put on a stove. If you rapidly change the temperature, you therefore also change the mirror's shape and focal length, even its ability to focus properly at all (yet another problem for refractors). Quartz and various ceramics have become popular, since they are not as sensitive to temperature variation. The 200-inch (5-meter) at Palomar is made of Corning Pyrex, which you can safely cook with. Whatever is used, since the reflective surface is on the front side, the backside of the mirror can have supports that keep it from sagging as it is moved around toward different parts of the sky. There is therefore much less limit on size, and thus on light-collecting power and on the attendant ability to resolve fine detail.

A reflector has an obvious problem. The focused image is in front of the telescope. To view it with an eyepiece, the observer must get in the way, blocking the view. To avoid the difficulty, Newton used a small, flat secondary mirror (rendering the objective the primary mirror) placed in front of the focus to send the light off at a right angle to the side of the telescope where it could be examined. The actual focal point without the secondary becomes the unused "prime focus." The secondary mirror does not put a hole in the field of view, but simply cuts down on the amount of available light, which one can compensate for by making a larger primary. The Newtonian focus, common in older professional reflectors, is still much used for smaller amateur instruments. The secondary mirror (be the telescope a Newtonian or some other design as described below) is usually mounted on an X-shaped strut, which causes its own diffraction pattern that shows up on bright stars as a cross, giving stars their "points." Refractors don't have secondary mirrors or struts to create such a pattern, so that if there is a bright star on a photograph, you can usually tell what kind of telescope the image was taken with.

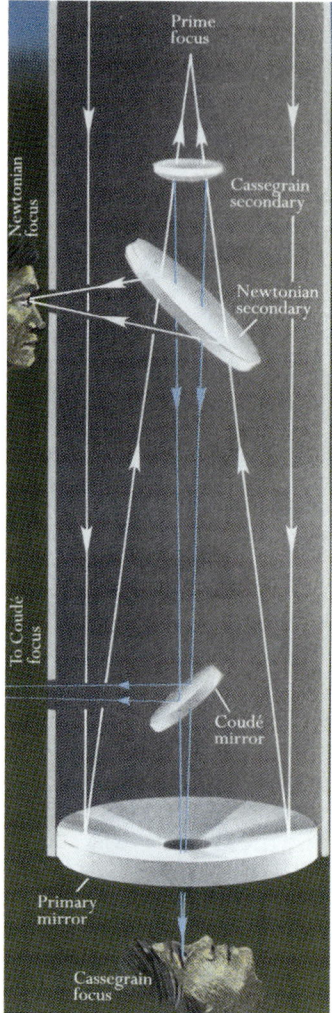

Figure 3.2. The curved (paraboloidal) objective mirror of a reflecting telescope brings parallel beams of starlight together at the prime focus. A tilted flat secondary mirror can send the light off to the Newtonian focus at the side (a popular amateur configuration). In the Cassegrain setup, a curved secondary mirror placed in front of the prime focus sends the converging beams back through a hole in the objective. An additional mirror in front of the objective can intercept the light from the secondary and bring it to a fixed "coudé" position. From "Stars," J. B. Kaler, Scientific American Library, Freeman, New York, 1992.

Reflectors have a great deal of flexibility. A common setup is not the Newtonian focus, but the Cassegrain, which was invented around 1670 and named after its presumed French inventor, M. Cassegrain (its history confusing at best). In a Cassegrain configuration, the secondary is a small convex mirror (shaped as a hyperboloid, another of the conic sections) set perpendicular to the objective's axis in front of the prime focus. The light is reflected back through a hole in the primary mirror where it is brought to a focus. The curvature on the secondary is designed to lower the angle at which the light beams converge, and thus effectively increases the focal length. You can get different effective focal lengths by changing the secondary mirror. The Cassegrain was preceded in concept by the rather similar Gregorian, designed by the Englishman James Gregory. It employs a concave, ellipsoidal secondary set beyond the prime focus, but then behaves pretty much the same way. Cassegrain, however, won. Gregorians are rare. But again, "Wait, there's more." By using additional mirrors, the light can be delivered almost any-where. A common setup is the "coudé," French for "elbow" (there is no M. Coudé; well there probably is, but he is not relevant here), in which the light is brought by multiple mirrors to a fixed point within the bowels of the observatory where it can be analyzed by equipment too massive to attach to the moving telescope.

There seems to be little physical limit as to telescope size. Herschel's largest instrument had a 48-inch mirror (bigger than the Yerkes objective), while in the 1840s, William Parsons, the third Earle of Rosse, built a 72-inch "behemoth" in Ireland. With it, he discovered the spiral arms in the nearby galaxy M 51, though no one at the time knew what galaxies actually were. His famous drawing of M 51 seems to have been an influence in van Gogh's "Starry Night." The 100-inch went up on Mount Wilson overlooking Los Angeles, California, in 1917, and remained the world's largest until 1948, when the 200-inch (5-meter) on Palomar Mountain went into ser-vice. The 100-inch was used to measure the first distances to galaxies, which led to the discovery of the expansion of the Universe. The 200-inch is so big that the observer can actually ride the instrument at prime focus (the passenger now replaced by electronics). The

Soviet Union upped Palomar a bit in the early 1970s by building a 6-meter telescope, but it turned out to be of limited usefulness.

The largest single mirrors are now around 8 meters (315 inches) across, more than half again the size of Palomar's great eye. All of them are in the American southwest, the Chilean Andes, or on Mauna Kea in Hawaii. Most impressive is the quartet of 8.3-meter (323-inch) instruments of the "Very Large Telescope" (VLT) of the European Southern Observatory in Chile. Even larger instruments can be made by using multiple mirrors to mimic the curvature of one big mirror, wherein the individuals are aligned by lasers. The grandest of these so far are the twin 10-meter (394-inch) Keck telescopes on Mauna Kea, operated by the Universities of California and Hawaii, Cal Tech, and NASA. The Kecks and the VLT telescopes can also be combined to use as one even larger instrument. Plans for 30-meter and even larger devices are underway. At this level, cost becomes the limiting factor. It was one reason for dropping the idea of the 100-meter "OWL" (the "OverWhelmingly Large telescope"; who says scientists have no sense of humor), one with a mirror as large as... a football field (see Chapter 2).

While refractors have their problems, so do reflectors. A parabolic mirror will bring light to a focus only on the optical axis. Away from the central position, images begin to spread out in an aberration called "coma" (because they look like little comets), which gets steadily worse toward the edges of the field of view. The various aberrations (there are others) can be corrected with more complex curves on the mirrors combined with the use of pre-focus lenses. Of particular interest is the Schmidt design (after the Estonian-German optician Bernhard Schmidt, 1879–1935), which consists of a curved correcting lens placed in front of a non-parabolic mirror and which allows good images over very large fields of view. The biggest of these, Palomar's 48-inch Schmidt, which also began service in 1948, has a 72-inch mirror coupled with a 48-inch corrector. With it (barring Herschel's vast work), astronomers made the first extensive survey of the sky, at least that portion of it seen from Palomar Observatory north of San Diego, California.

Directions please

Stars, planets, and other celestial objects are positioned on a coordinate grid much like latitude and longitude on Earth. The angle north or south of the celestial equator in the sky is the called the declination (see Chapter 2). That of the equator is 0 degrees, of the north celestial pole 90 degrees north, of the summer solstice 23.4 degrees north, etc. Longitude on Earth is measured east and west of the prime meridian that by common agreement runs through the Greenwich Observatory near London. The celestial analogue, right ascension (a historically-derived term; there is no left ascension), is measured to the east of the Vernal Equinox (again see Chapter 2). As we on Earth have our own personal latitudes and longitudes, so celestial objects have their own right ascensions and declinations. They are available from catalogues or, in the case of Solar System objects, through computation. From our perspective, the sky is not stationary, but turns around the poles as a result of terrestrial rotation. The location of the Sun relative to the celestial meridian (the north-south line through the celestial poles that divides the sky into its eastern and western hemispheres) gives the time of day, specifically the obvious solar time, a variation on which we keep as the standard familiar clock time. We can also define a sidereal (star) time given by the position of the Vernal Equinox relative to the meridian. The two kinds of time are related through the date (the time of year) and told by solar and sidereal clocks. All observatories have both kinds. The clocks are set by signals sent in several ways by a variety of government services that include the Global Positioning System (GPS).

The telescope is mounted on dual axes that are set perpendicular to each other. Traditionally, mostly for older classic instruments, rotation around one axis (the declination axis) moves the telescope north or south to point to a specific declination. The other, the polar axis, points to the celestial pole and is responsible for moving the telescope east or west by an angle computed from the right ascension and the sidereal time. Once the telescope is tilted to its target, it can be driven to the west by a motor synchronized to a clock to follow an object along its apparent daily path as the Earth rotates. Modern large telescopes have axes set up on the horizon and zenith (the point overhead), which provides much greater mechanical stability than one aligned to the celestial pole. The drive mechanism becomes more complicated since two motions are required at the same time, but the problem is easily solved by the telescope's computer. Telescopes can be pointed with remarkable precision, to within a fraction of a second of arc, which includes the lofting effect of atmospheric refraction.

Instruments, analysis, and whatnot

Not only is it fun to do, four centuries after Galileo, visual observation through the eyepiece still has a place in discovery. Comet hunters scan the skies at twilight and dawn for the tails of the icy remnants of planet formation, while other observers watch the brightness changes of variable stars or hop from one galaxy to another looking for stellar explosions. For many it's a relaxing, often deeply moving, hobby. But it's a niche market. Instead, research telescopes (and for that matter an increasing number of those belonging to amateurs) come equipped with a variety of detectors that can be mounted at the focus for permanent recording or analysis.

The simplest device is the camera, which holds a recorder of some sort and may or may not have its own lens (or mirror) system. The first permanent recording device, used from the late nineteenth century until near the end of the twentieth, was the photographic plate, which is set into a frame (the plateholder) mounted at the focal plane. The plate consists of a light-sensitive emulsion coated onto a glass plate, or onto a flexible substrate to make the once-familiar photographic film. Development in a chemical bath then brings out a granular negative image. Though positive prints are used widely for public delight and instruction, astronomers generally worked with the negatives. Not only was photography used to acquire images of celestial objects and position them precisely, it could also be used to estimate stellar brightnesses (magnitudes: see chapter 4), even if poorly. Best yet, photographic plates can be stored for future use, which becomes especially important for ephemeral events. The huge plate collections at Harvard and elsewhere are still mined for data.

At the most sophisticated level of photography is the great Palomar Sky Survey, which used the 48-inch Schmidt to photograph the entire sky as seen from southern California down to a declination of 33 degrees south on 935 pairs of 35-centimeter-wide glass plates through both blue and red filters. The survey was distributed (sold for $2000) world-wide on negative prints that looked nearly like the original glass plates, and was responsible for vast numbers of discoveries of celestial objects. The survey was then repeated from the southern hemisphere by the European Southern Observatory and distributed on negative transparencies. Both have been fully digitized and are now available from anywhere.

The chief problem with photography is its very low efficiency. It's not so noticeable in ordinary picture-taking, but it becomes crucial when you have only a few photons to work with. For every 100 photons, perhaps only one or two actually got recorded. And that was at the peak of the technology as developed (pun not intended, but pretty good anyway) by Eastman Kodak, a company that was revered by astronomers for its pioneering research into ever better emulsions. Nearly as bad, the photographic plate is non-linear. That is, doubling the brightness of the infalling light does not produce a doubling of the darkening of the emulsion, but sometimes less, sometimes more, so the plates had to be calibrated simultaneously with the observations, adding another (rather large) error-inducing step, which made it devilishly hard to determine accurate relative stellar brightnesses. Moreover, photographic emulsions are sensitive primarily to blue (and shorter-wave) light. Taking images at red wavelengths was more difficult (though done, as in the Palomar Sky Survey), and in the low energy infrared nearly impossible.

A new era began with the first electronic device, developed around 1912 at the University of Illinois Observatory in Urbana (rather like Hal in "2001: A Space Odyssey"). It used a light-sensitive selenium cell at the focus of what is now a historic 12-inch refractor to produce an electrical signal that responded to starlight. Though at first photoelectric photometers were limited to observing one star at a time, the sensitivity of the device to stellar brightnesses was vastly better and closely linear. Moreover, in more modern form, it could be used at longer wavelengths.

Photography for wider-scale imagery and measure was first replaced by video systems and then by the "charge coupled device," the CCD, which records electronically for playback by computer and is basically the same thing that is used in digital cameras except for being larger and more precise. At peak wavelength sensitivity, CCD photon efficiency approaches 100 percent. Exposures that took hours with traditional photography can now be done in minutes, even seconds. Moreover, not only can a field of view be quickly imaged, but the brightnesses and locations of huge numbers of objects, from stars to galaxies, can be measured all at the same time. Moreover still, the CCD is also linear: double the brightness and you double the electric, recordable, charge. The CCD opened the door to floods of data that can be used immediately or stored for future use in a "virtual observatory." Telescope systems are now being designed to image and measure the entire sky over only a few days to unheard of faintness limits far below that of the Palomar Sky Survey, which took years to complete. Comparison of millions of objects from one night to the next makes it possible to discover and monitor hosts of variable stars, sudden explosive events, even passing solar system objects such as rogue (and even ordinary) asteroids. Such systems are possible only with the aid of highly advanced computers by which astronomers can control the telescope and store the data. But by holding a record of the past, photography still has its place, as do ancient visual observations, without which we would know little of a great number of historic events.

Other than imagery, which is closely bound with brightness measurement, the most important instrument (some would say the most important) is the spectrograph ("spectroscope" when used visually). Its obvious purpose is to spread the light out into a spectrum for the determination of the strength of radiation at various wavelengths and for the measure of the positions and strengths of absorption or emission lines (Chapters 1, 5, and 6). The first spectroscopes and spectrographs used prisms to form spectra. Unfortunately, prismatic spectra are not linear with wavelength, the spread, or dispersion, of starlight increasing with decreasing wavelength, making wavelength measures difficult. A modern spectrograph uses a diffraction grating, a closely grooved plate that creates spectra through the interference of light waves. Diffraction spectra can be seen in light reflected from the surface of a compact disc (see Chapter 1). They can be vastly stretched to yield great detail. Moreover the dispersion is linear with wavelength, greatly simplifying the reduction of the observations into useable form.

Various devices that go back to the 1890s are available to record multiple spectra, vastly increasing the efficiency of the telescope. The detectors remain the same, beginning with the photographic plate then culminating in the electronic CCD. The best known of these is the objective prism spectrograph used by the Harvard observers in the creation of the early twentieth century's Henry Draper Catalogue of stellar spectral classes (Chapter 6). Instead of putting the prism at the focus of the telescope, a large, fairly flat prism was placed over the objective so that images of stars were replaced by images of their spectra. Needless to say such devices were difficult to use in the Milky Way. Modern multiple-imaging spectrographs can get hundreds of accurately calibrated spectra at the same time.

Domes and mountains: The observatory

The cartoonist has the telescope sticking out of a slit in a dome. It's enough to make an astronomer cry. The dome, with its famous slit shuttered when the telescope inside is not in use, provides weather protection. The dome rotates so as to place the open slit in the right direction to accommodate the telescope. A variety of other more modern housings are used as well, including at the extreme nothing at all.

In days of old, in times of difficult travel, observatories were generally built where the astronomers lived. You'll find refractors and modest reflectors scattered all over the eastern United States. Owned by the University of Chicago and no longer used for research, Yerkes Observatory with its 40-inch refractor is just 90 miles from the big city. Bright lights and thus bright skies have made such locations impractical if not impossible to use. Not only that, at but a few hundred feet above sea level there is a lot of light-absorbing air that the old telescopes had to look through. Not to mention a lot of cloudy skies.

Among the worst problems for any ground-based telescope is the twinkling. The air (Chapter 1) is a refracting medium that even disperses starlight and moreover is hardly a uniform sheet. It consists of cells of different densities that refract and disperse through slightly different angles and blow about. The path taken by starlight to the Earth then zigs and zags as it passes through the cells. An observer on the ground sees the star erratically jump about, change colors, and brighten and dim. It can be a pretty sight, "twinkle twinkle little star" and all that, but it makes a mess of the telescopic image, which heaves about, creating a boiling colorful "seeing disk" that might be a few seconds of arc across, one that can easily dwarf the diffraction disk and destroy the instrument's resolving power.

It's far better to place an observatory in a remote, dark site far from city lights and on a mountain top as high as possible, where the starlight is much less impeded such that the twinkling settles down (though it never goes away). As important, more light (especially shorter-wave blue, violet, and ultraviolet) gets through to the ground. Given modern transportation, observatories can be located far from the astronomers' homes. The mountains of the American southwest provide fine locations, as do the Andes of South America, where observatories seem to litter the landscape. Perhaps the best site on Earth is Mauna Kea, a 13,800-foot (4200-meter) dormant (but perhaps rather frighteningly not yet dead) volcano on the Big Island of Hawaii, where seemingly vast numbers of telescopes patrol the skies. Here the seeing disk may be half a second of arc across or even under, yielding exquisite views. As important, at a properly selected site, the sky is predictably clear.

At a modern observatory, it's not even necessary to have the astronomer present. With computers and video systems, the observers can sit comfortably thousands of miles away from the actual action, as they must be while using a telescope in space. At the extreme, the requested observations are put into a queue along with others and the data shipped back to the office, whether by telecommunication channels or by just flying recording devices back and forth. We have, however, lost the delight of working in a freezing dome or battling bugs while braving the dangers of working with massive equipment in the dark (astronomy was once listed as among the more dangerous of occupations). But at least we were under the stars where we could glory in the artistry of the nighttime sky and the beauty of the Milky Way.

And then came hubble

Twinkling, the matter of the seeing disk, is troublesome. Even more so is the problem of atmospheric absorption, which blocks most of the ultraviolet, much of the infrared, the X-ray domain (fortunately for life), and various radio (see below) bands. The best solution to the atmospheric problem is to eliminate the air altogether by putting telescopes entirely above it. There have been a large number of different kinds. The best known is the general-purpose Hubble Space Telescope (HST).

The idea is not new, but goes back to 1922 and the German rocket scientist and futurist Hermann Oberth. It was conceived in more modern form in 1946 by Cornell's Lyman Spitzer. Unfortunately, the concept had to languish until the Soviet Union and then the US could launch and control ever-larger satellites. The original plan was for a three-meter orbiting telescope, which was eventually scaled back to 2.4 meters (94 inches) to fit the Space Shuttle bay that had, with a lot of grumbling, been adopted as the launch vehicle. Delayed by the Challenger explosion, the HST, named after Edwin P. Hubble (who discovered the expansion of the Universe), was finally launched in 1990 into an orbit 350 miles (570 km) high. The HST, run by the Space Telescope Science Institute at Johns Hopkins University in Maryland, came with a full package of instrumentation, which included cameras, spectrographs, and the ability to measure precise position. All instruments were available for use by rotating them on and off the focal plane.

Figure 3.3. The Hubble Space Telescope orbits the Earth above our disturbing atmosphere, allowing high resolution of detail as well as observation of the ultraviolet and infrared parts of the spectrum. The 2.4-meter (94.5-inch) mirror is positioned at the right end, where it is accompanied by a suite of imagers and spectrographs. Launched by the Space Shuttle in 1990, five servicing missions have kept it going. Space Telescope Science Institute.

Designed to resolve detail to only a few hundredths of a second of arc, to the astronomers' initial horror the HST could not be brought to a proper focus. The mirror had not been ground and polished to the correct curvature. Now we could only be thankful that the telescope had been married to the Shuttle, because the craft could then be serviced. In one of the most amazing space adventures of all time, in 1993 astronauts went back to the telescope to install corrective optics. They worked perfectly. Three more visits to the telescope over a period of a couple decades delivered additional, new, and better instruments and repaired the electrical support and pointing systems. Thanks to the release of hundreds of carefully processed color images and an excellent and resourceful public education program, Hubble became the "peoples' telescope." In the face of a decision by NASA to let a failing instrument go, the public demanded a last, fifth, visit that would allow it to continue to send back superb science data and captivating imagery well into the 21st century before it is finally put to rest, de-orbited safely into the ocean.

Back home, astronomers and engineers were making ground-based observatories competitive with "adaptive optics" that allowed the de-twinkling of starlight. To obtain a steady image of some celestial object, the telescope monitors a star within the field of view and follows its rapid atmospheric meanderings. The optics of the telescope are then almost instantly distorted via small motors to compensate. The result is a stilled image that approaches what was seen from space. The problem with this approach is that there may not be a suitable star to follow within the chosen field of view. So we make one with a laser beam focused on the upper atmosphere to create an artificial star that twinkles like any real one, with the same result. All modern telescopes are equipped with such adaptive optics systems. The Hubble, however, has another spectacular advantage. Not only does it not need a guide star, but it can look in wavelength bands in the ultraviolet and infrared that are inaccessible from the ground, expanding the electromagnetic spectrum to observation, a feat that had actually been started long before, but in a different way.

Opening up

One of the great astronomical achievements of the twentieth century was the opening of the full electromagnetic spectrum to observation. The first adventure took place at long wavelengths. In 1933, Karl Jansky, an engineer working for Bell Labs, was assigned to track down the source of the background radio radiation that was interfering with transatlantic radio communication. He set up a radio antenna in Holmdale, New Jersey, that looked like a part of a giant bedspring. An antenna in its simplest form is no more than a wire or a rod in which an electrical signal is excited by passing radio waves of various wavelengths or frequencies. The signal is then electronically isolated (tuned) to a specific frequency, amplified, and recorded. A simple rod (or dipole) antenna is more sensitive in the perpendicular direction than along its length, and is thus crudely directional. With a more complex antenna and better directionality, Jansky found that the background was strongest when the Milky Way went through the most sensitive direction of what was now, though he did not know it, the world's first radio telescope. A few years later, an ingenious American amateur scientist named Grote Reber (1911–2002) hand-built the first dedicated radio telescope with a more familiar design, one that used a 31-foot (9.6-meter) parabolic dish (a radio mirror) to focus radio waves onto an antenna (the detector) placed at prime focus, with which he made the first map of radio radiation from the Milky Way. It worked just like an optical reflector, only at longer wavelengths, his first successful observation made at a wavelength of 1.9 meters.

The field of radio astronomy had to wait out apathy (astronomers did not know what to do with radio waves from space) and World War II, after which it exploded in part as a result of the advance of wartime technology. Radio telescopes are now all over the globe, searching and recording the radio sky in a vast array of wavelength bands to which the Earth's atmosphere is transparent, many of them protected by world agreement against encroachment by artificial signals. (Many too are the radio bands that are blocked by the air, for example those used for short-wave radio in which the Earth's ionosphere reflects signals over long distances. These remain closed to astronomers, as the size of radio telescopes makes it difficult to place them into orbit, though it has been done.)

It's vital to note that radio astronomers do not "listen" to radio waves from space. There is no audio signal, no "modulation." In what we think of as "radio," commercial or otherwise, an audio signal is piled aboard the radio wave. The radio receiver's job is to pick up the modulated wave (AM for modulation in amplitude, the size of the wave; FM for modulation of frequency), then throw it out and keep the audio signal. Instead, astronomers "look" at the radio waves themselves (the radio "noise" if you will), measure their intensities at different locations on the sky, and create radio maps, even computer-generated pictures of things at different wavelengths that can appear just like those made with optical telescopes.

As in the optical domain, radio telescopes come in a wide variety of designs that include the Cassegrain, with the focus behind the parabolic dish. Some, like Jansky's original, don't even look like telescopes. Radio telescopes also come equipped with spectrographs to record both emission and absorption lines created by a variety of astronomical objects across a huge range of wavelengths, from submillimeter on up. A home radio, whether digital or analog, is a radio spectrograph that is tuned across the spectrum to pick up radio stations that broadcast at specific frequencies and are quite literally emission lines. With radio radiation we can pick up signals from processes that radiate little or nothing in the optical and that we never knew were there, especially those with low internal energies. We can probe into dark, cold interstellar clouds to observe star birth in action, chart out the chemistry of interstellar space, even examine the magnetic fields of nearby planets. The list is endless.

Radio telescopes have one huge problem. The size of a telescope's diffraction disk (the image of a point source) is, as noted before, inversely proportional to the telescope's aperture; but it is also directly proportional to the wavelength being observed. For an optical telescope, the wavelengths are very small, hence so is the disk, which gives optical instruments their great ability to resolve detail. Radio wavelengths, however, are so long that the resolution of detail is awful. To compensate, radio reflectors must be made proportionately larger than optical instruments. And therein lies the problem. Simply compare one with the human eye, which has a maximum aperture of about 7 millimeters and a resolving power of at best a minute of arc. To achieve just ordinary vision working at a wavelength of 10 centimeters requires a telescope roughly 1300 meters, or 1.3 kilometers (0.8 miles), in diameter. Needless to say, a conventional telescope of such proportions is an engineering impossibility. We can cut the size to a more reasonable (but still huge) 130 meters by observing at 1 centimeter, but nature dictates the wavelengths to be observed, not desire. The largest fully steerable radio telescopes have parabolic radio mirrors 100 meters in diameter, the size of the ubiquitous (here it is again) football field, which achieves human resolution at wavelengths just under a centimeter. But the most important radio emission line from space is more than 20 times that long (the 21-centimeter line of neutral hydrogen, used to map our galaxy as well as others). In the 1960s, the Navy attempted a 600-foot dish in West Virginia (an interesting place for a Navy) for communications use and gave up (to be fair, in part for reasons of technology that were advancing in other areas).

With conventional means not feasible, we must proceed to the unconventional. A radio telescope does not have to be fully steerable. The Arecibo radio reflector, 300 meters (1000 feet) across, is fixed and set into a natural depression in the hills near Arecibo, Puerto Rico. It's directed straight up and is "steered" east and west by the Earth's rotation (observing astronomical objects as they cross the celestial meridian), and can also be directed a bit north and south of the zenith by shifting the receiving antennas. The Arecibo facility is also the world's most powerful radar, and has been used to examine the planets out as far as Saturn. In radar (originally an acronym for "radio detection and ranging"), a signal is beamed out to the object and the reflected signal analyzed in a variety of ways. It's most familiar in aircraft and traffic control. The best way of calibrating the Astronomical Unit (the average distance to the Sun) is by measuring the time it takes a radio signal to get to Venus and back at the speed of light. The distance of Venus is known precisely from orbital theory in Astronomical Units, while the radar observation gives it to us in kilometers, thus telling how many kilometers there are to the AU. Radar can even provide planetary images and rotation rates, as was first done for Venus in 1962.

But no matter how hard astronomers might try, they could not pick up the Arecibo dish and point it around the sky, so we are still position-limited to a few degrees of declination off the zenith. We can, however, perform an amazing trick. Point two ordinary radio telescopes toward the same target and mix the incoming signals together via cable. The waves interfere with one another, producing highs and lows as the source moves through the field of view (rather like the diffraction rings of an optical telescope). With appropriate mathematics the "interferometer" has the resolving power (though not the collecting area) of a single dish with an aperture equal to the

Figure 3.4. Working like an optical telescope, a traditional radio telescope uses a parabolic reflector to send radio waves to a detector placed at the focus. This particular instrument is part of the Very Large Array, or VLA, in New Mexico, which can mimic the resolving effect of a telescope equal in size to the smaller telescopes' physical separations. Two other members of the array can be seen at lower right. J. B. Kaler.

separation. With just two radio telescopes there will be an inevitable loss in detail. We can overcome that defect by adding more individual radio telescopes to fill in the space that would be taken by one big dish, allowing us to mimic its effect rather nicely. Interferometers have been around since the 1950s. The best known is the Very Large Array (the VLA, go visit) that covers 27 miles of desert in New Mexico and can imitate the effect of a single telescope of that size.

We can do even better. Using precise timings instead of direct cable connections, Very Long Baseline Interferometers (VLBIs) have been made roughly as large as the Earth, which gives an amazing ability to sense fine detail, one far greater than is possible with conventional optical telescopes, turning the matter of resolution on its head. It's now possible to achieve resolving powers the order of several millionths of a second of arc across a large span of the electromagnetic spectrum. Similar techniques have been extended to shorter wavelengths, giving the optical domain a chance to fight back, though the technologies are still in their infancy.

Onward

Between the radio and optical domains lies the vast infrared, which starts at wavelengths somewhere below a small fraction of a millimeter (there is no strict definition) and ends where the human eye picks up at around 0.000075 centimeters, or 7500 Ångstroms (Chapter 1). While infrared radiation was discovered by Herschel in 1802 (Chapter 1), detector technology was slow to develop, though now infrared detectors are as good as those in the optical. Even so, many wavelength bands are blocked by water vapor and carbon dioxide in the Earth's atmosphere, which allows only a limited view from the ground. (Absorption of radiation from the Earth by these same bands provides the greenhouse effect that helps keep the Earth warm enough for life and also causes global warming as a result of carbon dioxide, methane, and other greenhouse gases being pumped into the air.)

To overcome this problem, we go back into space. While the Hubble Space Telescope can cover the near infrared (that part closer to the optical domain), it's limited in its spectral span. Several other orbiters, however, have been dedicated to the infrared. Among them was the highly effective Infrared Astronomical Satellite (IRAS), which in 1983 observed the sky in four wavelength bands from Earth orbit and catalogued more than a quarter million infrared sources. It was followed by the versatile Spitzer Space Telescope (launched in 2003 and named after one of the promoters of the HST; see above) and the Herschel Space Observatory, which went into service in 2009 and covers the far infrared and submillimeter, beyond which we call the radiation "radio." The six-meter Next Generation (now James Webb) Space Telescope, scheduled to replace Hubble, will be optimized for the infrared so as to study the expanding Universe.

As effective as they are, infrared space observatories have their own problems, as they must be cooled to near absolute zero to protect their detectors from being overwhelmed by solar heating and infrared radiation from the observatories themselves. Once the coolant (liquid helium) runs out, the infrared telescope becomes much less effective.

Going to shorter wavelengths, we zip through the optical to the ultraviolet. An important feature of Chapter 1 is that only a small bit of the near ultraviolet gets through to the ground (its rather dangerous high energy producing sunburns). The rest, and downward into the vaguely defined X-ray domains, are (fortunately for us) again blocked by the Earth's atmosphere. A number of orbiting telescopes launched to study high-energy processes have attended to it, including the International Ultraviolet Explorer (an amazingly long-lasting ultraviolet spectrograph), the obvious Extreme Ultraviolet Explorer, the X-ray's CHANDRA, and the Fermi Gamma-ray Telescope. Needless to say, the whole spectrum has been covered with increasing effectiveness, and there is more to come.

To this vast array of instrumentation, add the Laser Interferometer Gravitational Wave Observatory, a massive device that is detecting the most distant and violent events in the Universe. Consisting of two interferometers each four kilometers across and separated by 3000 kilometers, LIGO looks nothing like a telescope at all. But then neither do the neutrino telescopes of Chapter 1 nor Jansky's first radio telescope. It's all in what they do, not in what they look like. With all this in mind, let's back up and look a bit at what our own eyes can see, knowing that "you can observe a lot by watching."

4

A Celestial Tour

In times past, the children of the aristocracy would take the grand tour of Europe to view the sights and help make them more worldly and sophisticated. Here we do the same with the sky, wherein we will more or less randomly romp among the stars and constellations to sample the celestial scenery, the idea really being to intrigue. We'll take various side excursions, after which we'll return to the main tour. In later chapters, we'll look at where the various celestial sights came from, how they fit together, what will happen to them, and ultimately what they mean to us.

Constellations: A continuing theme

A typical autumn evening view from the northern hemisphere shows the stars of the Greater Bear treading down the northwestern sky. Its Latin name, used also in English, is Ursa Major ("ursus" is Latin for "bear"). Does it look like a bear? Sure it does. If you relax the imagination there are stars that mark the tail, the body, the snout, and three feet (hey, nobody's perfect). It's a prime example of a constellation, a pattern of stars to which somebody long ago gave a name. Most cultures seem to have made them up, from China to native America. Our "western" ones come rather oddly out of the middle east, out of Mesopotamia, then found their way to old Greece, their earliest mention around 800 BC. They were later helped along, saved really, by the ancient Arabs, then sent back to us today. Painted on the sky, they tell vibrant stories and important cultural myths. The patterns do not necessarily look like what they are supposed to be. No artist created them. They represent, not portray, though a few are really quite obvious. Ursa Major is a formal constellation, one of the 88 adopted worldwide from ancient and more modern lore by the International Astronomical Union around 1930 to partition the sky into smaller sections, mostly to aid in the naming of stars and other celestial objects. It's remarkable that a concept thousands of years old is still in use. That hardly means that the celestial patterns of other cultures are gone; they just add to the mix.

Within Ursa Major is one of the most beloved of stellar patterns, the Big Dipper, a figure that truly does look like what it's named for. You almost get the sense that when it dips under the pole, compliments of the Earth's rotation, it scoops up distant waters to refresh the admiring viewer as it rises in the northeast. The Big Dipper is a fine example of an "asterism," an informal constellation, many of which dot the sky. Other cultures see it differently. To the English, the Dipper is the Plow, or Wagon, while to the ancient Arabs it was a funeral bier with the daughters (the tail of our Bear) following behind.

Figure 4.1. The obvious pattern of the Big Dipper shines through the slit of the University of Arizona's 90-inch telescope on Kitt Peak. The Dipper is the tail and backside of the constellation Ursa Major, the Greater Bear. J. B. Kaler.

To see the Smaller Bear, follow the two front bowl stars of the Big Dipper (the "Pointers") to Polaris, the North Star, which by accident sits within a degree of the North Celestial Pole (Chapter 2). Polaris also lies at the end of the handle of the fainter Little Dipper, which forms the tail and body of Ursa Minor. Look more closely at the tails (Major's bent down, Minor's bent upward). Have you ever seen a bear? There is no long wagging tail. How do you get such a beast into the sky? You very carefully sneak up behind it and quickly grab its short tail. Whirling the bear around your head, you throw it upward to the heavens, which stretches out the tail. Sitting around the campfire after a long day of herding, farming, or gathering, your tribe's storyteller could stretch out that tale (sorry) until the campfire died and the potables were consumed. Such is the nature of the sky, which becomes both a secular and sacred storyboard. We are no different. What's the campfire today? Television or some other device around which we collectively sit and still hear the old stories that never fade but just appear in new guises.

Forty-eight constellations descend to us from the ancient Greeks through the poets and finally from the second century Alexandrian astronomer Ptolemy. The division of Argo (the Ship of the Argonauts that floated across the Greeks' southern skies) into three parts in the eighteenth century gives us 50. The ancients, however, could not see the stars of the deeper southern hemisphere. Moreover, there were lots of spaces between the ancient constellations that held only faint stars and that made no obvious patterns. In the seventeenth and eighteenth centuries, the old sky figures were supplemented by numerous modern (how time flies) constellations created to fill in the blanks, especially within the large area of the sky surrounding the South Celestial Pole. From a wealth of patterns that included real and fanciful animals, as well as artifacts of the new times, 38 were finally adopted along with formal horizontal and vertical boundaries, giving us an orderly 88. To these we add the seemingly endless number of informal figures, of asterisms like the Big and Little Dippers, as well as those that were ultimately, and perhaps mercifully, rejected (Bufo, the Toad, comes to mind). You can make some up yourself and see them as old friends for a lifetime.

Star light, star bright

Several matters now arise. Although the question might seem a bit silly (presumably we know one when we see one), how do you define a star? Return to the Sun, which is supported by the fusion in its core of hydrogen into helium with loss of mass and gain of energy. A star might best be defined as a body that runs on core fusion, once did (to account for dead stars), or will in the future (to include forming stars). Or, as we will see, not.

Among observables, brightness comes first. Whether you know your constellations or not, it's obvious that while they are prominent, the Dipper stars are not the brightest in the sky. Many surmount them. They are mostly of "second magnitude," as is Polaris at the end of the handle of the Little Dipper (a popular misconception erroneously having it as brightest of all stars). In the second century BC, Hipparchus of Nicea, perhaps the greatest of the ancient astronomers, divided the sky's stars into six bins according to apparent brightness, or "apparent magnitude," first magnitude the brightest of them, sixth the faintest he could see. The magnitude system is still in use today, though modified and mathematically defined. There are 22 first magnitude stars. Fainter stars are vastly the more numerous. Going to deep sixth we can count a total of around 9000 that are potentially visible to the naked eye depending on one's visual acuity. To help further visualize the system, the faintest of the Big Dipper stars, Megrez, placed where the handle joins the bowl, is third magnitude. The Little Dipper provides a better example. In addition to Polaris, the front bowl star (Kochab) is also second magnitude, the lower front bowl star (Pherkad) is third, while the other three stars of the handle are fourth, and the faintest of the bowl stars is fifth.

Like the audio scale of decibels, magnitudes are logarithmic (stay with me…), in which small numbers signify a big change. In the nineteenth century, the system was formally set up such that five magnitude divisions relates to a factor of 100 in visual brightness. First magnitude stars are on average 100 times brighter than those of sixth magnitude. One full magnitude difference then corresponds to a brightness factor of the fifth root of 100, or 2.512…, two magnitudes to 2.512 ×

2.512 or 6.3…, 10 magnitudes to 100×100 or 10,000, and so on. To scale the system, which is continuous and decimalized, Polaris, visible from all over the northern hemisphere, was originally set at magnitude 2.0. (Current calibration with other stars makes it 2.02; Polaris is also slightly variable, not a good thing for a standard.) Dubhe, the front bowl star of the Big Dipper, has a modern magnitude of 1.79, while the faintest star of the Little Dipper gleams weakly at 4.95.

When we so calibrate stellar brightnesses, the very brightest of stars exceed first magnitude and extend to magnitude zero or even -1, the actual brightest star, Sirius, shining at -1.46. Other celestial objects can be much brighter. At its most luminous, Jupiter glows at -3, Venus at -5, the full Moon at -13, the Sun at a dangerous -27. Generic first magnitude extends from 0.51 to 1.50, second from 1.51 to 2.50, magnitude zero from 0.50 to -0.49 etc. (though those few of magnitude zero and -1 are still casually lumped into "first magnitude"). With optical aid we can go much fainter. Binoculars can take the view to eighth magnitude or better, a small telescope under a dark sky to tenth or fainter (larger numbers always signifying fainter stars). At best, with Hubble or even from the ground, telescopes can detect stars as faint as 30, four billion times fainter than the dimmest that can be seen with the unaided human eye. It's also now possible to measure magnitudes to accuracies well under a thousandth of a division, which allows subtle stellar variability to be measured.

Traditional magnitudes are a measure of brightness as seen with the human eye, which has its peak sensitivity in the yellow-green part of the spectrum, just where the Sun shines at its brightest (hardly a coincidence). Those defined by Hipparchus (and later more technically) are thus "visual magnitudes." But different stars with different temperatures shine at their best at different wavelengths (Chapter 1). A cool star (say a few thousand degrees) will radiate mostly in the low energy red part of the spectrum, a cooler one in the infrared, a hot one (maybe 20,000 Kelvin) mostly in the short-wave energetic violet, even ultraviolet, and so on. The magnitude thereby depends on the wavelength used. An "eye" sensitive to blue light will see hot stars the brightest, while one sensitive to red will see the cool stars best. Some very cool or dust-shrouded stars are entirely invisible at visual wavelengths, yet glow brightly in long wave spectral bands. So astronomers use magnitudes as seen at different colors, blue magnitudes, ultraviolet magnitudes, red, infrared, and so on unto enormous complexity. Differences between magnitudes as measured in different colors (called "color indices") are sensitive to temperature, and thus become a fairly accurate quick proxy for it, as temperature can be time-consuming to calculate directly from the spectrum. The standard magnitude, however, will remain the good old visual variety, properly called "apparent visual magnitude."

Far out, man

On a dark unstormy night, deep in the country where you can get away from artificial lights and there is no Moon, with literally thousands of stars gleaming overhead, a child might be tempted to run to the top of a hill to pluck one down. It won't work. The stars are so far away that the first distance was not measured until 1838, when the German astronomer Friedrich Bessel (1784–1846) detected the parallax, the tiny shift in stellar position seen as the Earth orbits the Sun, of 61 Cygni. You get the same effect by looking at a nearby object with one eye then the other. The farther the star, the less the shift, which at best is just over a second of arc.

Even the nearest of stars is unimaginably distant. It takes light, at a speed of 299,800 kilometers per second (186,300 miles, more than seven times the circumference of Earth), just 1.3 seconds to traverse the distance from here to the Moon, 8.3 minutes for a photon to make the journey to us from the Sun. A signal from a spacecraft orbiting Pluto takes some four hours, which (as noted) would make for a difficult conversation were there an astronaut aboard. Then the gap hugely widens. Distances to stars are measured in light years. The light year is the distance that a light ray travels in a year of 31.56 million seconds ($365.2422 \times 24 \times 60 \times 60$). Multiply that by the speed of light, and you get a distance for the light year of 9.46 trillion kilometers, 5.88 trillion miles, or 63,240 Astronomical Units (coincidentally about the same number of inches in a mile). The nearest star (Alpha Centauri) is 4.32 light years away, 41 trillion kilometers,

25 trillion miles, or 273,000 AU (Bessel's 61 Cygni at 11.4 light years). No, you can't pluck one down. Voyager 2, the spacecraft launched in 1977 to visit the outer planets, is now over 130 AU, nearly 19 light hours, away. If directed at Alpha Centauri (which it isn't), at its current speed of 16 kilometers per second (which is slowly lessening thanks to solar gravity) it would take some 80,000 years to get there. Humans have been to the Moon, 0.0026 AU away. Star travel in the near (or even distant) future does not seem likely. And that's the closest one. The most distant stars seen with the naked eye are thousands of light years away.

The distances are so large that light travel time becomes a bit of an issue. If stars are that far away, Alpha Centauri, Vega, or Deneb could blow up (the first two can't, the third will: more about that later) and you would not know it for tens, even thousands, of years. That's true. We see Alpha Centauri as it was more than four years ago, not as it is "today." But we also see the Sun as it was eight minutes ago, and the Moon as it was just over a second ago. Nobody cares. Even if one did, there is nothing to be done about it. Make it personal. It takes light about trillionth of a second to go a foot (30.5 centimeters). You therefore see the people around you as they were a few trillionths of a second ago. It doesn't matter or bother anybody. The Universe, from people to stars, exists as we see it. Everything is thus seen at a different time, while different people must have different views than you. Here is one of the foundations of relativity: there are no absolute views of anything.

If we have distance, we can distinguish between apparent and absolute brightness, which relates to actual wattage. The Sun, for example, radiates at a rate of four hundred trillion trillion watts (Chapter 1). Instead of watts, astronomers begin with the more arcane concept (no surprise) of "absolute magnitude," the apparent magnitude (that actually seen by eye) that the star would have at a standard distance of 32.6 light years. Eh? At a distance of 3.26 light years, the radius of the Earth's orbit (the AU) would appear to be a second of arc across, and the star would be one "parsec" (a common distance unit) away. The standard for absolute magnitude was arbitrarily set at 10 parsecs (or 32.6 light years) in the nineteenth century.

The apparent brightness of a light depends inversely on the square of its distance as the light spreads over the surface of a sphere. Put it 10 times farther away, and it will appear a hundredth as bright. Armed with distance, the absolute magnitude can then easily be calculated from the apparent magnitude. If all the stars were at the standard distance, the sky would look terribly different. Vega (second brightest in the northern hemisphere) would appear about the same, but first magnitude Deneb would shine at magnitude −7. Several times brighter than Venus, it would cast modest shadows and be easily visible in a clear blue sky. The mighty Sun, however, if so placed would shrink to fifth magnitude and glow feebly at about the apparent brightness of the faintest star in the Little Dipper. With proper scaling, absolute magnitude in the visual band of the electromagnetic spectrum, along with correction for invisible infrared and ultraviolet light (which depends on temperature), then allows the full wattage, or true luminosity, to be known. The range is remarkable. At the lower limit, we find stars (depending on definition) with luminosities under a ten-thousandth that of the Sun. Some are so visually faint that if you put them at the edge of the planetary system, they would not be visible at all (though we could detect their infrared signatures). At the high end, they can radiate at millions of times the solar luminosity. Exploding stars are even brighter. Surface temperatures equally astound, running from near that of a self-cleaning oven to more than 100,000 Kelvin, the Sun rather nicely in the middle. Dim stars are by far the most common, the most luminous ones are thankfully (they are dangerous) very rare.

The galaxy

All these stars and more than 200 billion others are part of our galaxy. Most of its stars, including the Sun, lie in a flat, rotating disk 100,000 light years across. We are set off to the side about 25,000 light years from the center, roughly half way out. But there are really no formal boundaries. The galaxy's true extent is much larger, the system just fading away into the dimness of space. All stars orbit the galaxy's center, the Sun taking some 200 million years to make a circuit.

Figure 4.2. Were we able to view our galaxy from afar, it might look something like Messier 83 (the "Southern Pinwheel") with its dramatic spiral arms set into a flat disk. If ours, the Sun would be somewhat over halfway out, from which vantage point the disk would appear as a surrounding Milky Way that is brightest toward the center. The red blobs are diffuse nebulae illuminated by the ultraviolet light from hot stars. They indicate regions of active star formation. At the center is a hidden supermassive black hole. NASA, ESA, and the Hubble Heritage Team (STScI/AURA).

We see the combined light of the disk's countless stars around our heads as the Milky Way, which was first resolved by Galileo. Lost among the bright lights of town, from the dark countryside the Milky Way can be a spectacular sight. From our off-center position in the galaxy, the Milky Way varies mightily in brightness from the direction of galaxy's center in the zodiacal constellation Sagittarius to its opposite in Taurus. Running down the mid-line of the Milky Way is a ragged dark rift caused by clouds of interstellar dust that embrace stellar birthplaces. Within the galaxy's disk is a set of spiral arms that wind outward from the center in which luminous young stars (and thus the dark birthing clouds) are concentrated. If we could get far outside the galaxy and watch, the arms would come and go with time, though with agonizing slowness. Surrounding the disk is a vast, sparsely-filled halo occupied by ancient stars that are the remains of the galaxy's original inhabitants. At dead center is a huge black hole with a mass four million times that of the Sun, whose gravity is so strong that even light cannot escape. It's made visible by the heated remnants of torn up stars circulating around it.

Our galaxy is but one of a vast, uncountable number of other galaxies, hundreds of billions of them, going off into distances that stretch billions of light years away. Some are shaped like ours with disks and spiral arms, while others are ellipsoidal blobs with little internal structure or active star formation. Scattered among the large galaxies like the Milky Way (and some much bigger) are great numbers of small, irregular dwarfs.

Galaxies tend to cluster. The "Local Group" of a few dozen galaxies (most of them dwarfs and scrappy irregulars) is dominated by ours and a similar system in the constellation Andromeda that is visible to the naked eye even at two million light years away (making it the farthest thing easily seen without optical aid). At the upper limit, some clusters contain thousands of galaxies. Between them all are scattered singles, doubles and multiples. Here and there are colliding systems, wherein larger galaxies are made from smaller ones. What we see is but a small fraction of the mass that is there, as galaxies seem to be embedded in some kind of mysterious, dominating "dark matter" that has a powerful gravitational effect but is otherwise completely invisible.

All but some galaxies in the Local Group are receding from us at speeds in direct proportion to their distances, which implies that the Universe is expanding. The expansion is most likely the result of a singular creation event, the Big Bang, which took place almost 14 billion years ago. Only among the most distant galaxies is light travel time important, as it allows us to look far back into the past to see what happened and how the Universe and its galaxies were born and raised. The concept of the Big Bang is powerfully supported by the three-Kelvin microwave background, which surrounds us all and is the remnant of the original Big Bang fireball. In the beginning, all the stuff of the Universe, at that point being just energy, was compacted into a hot dense state. As the Universe expanded and cooled, energy converted to matter, out of which primitive galaxies quickly condensed, perhaps around hugely massive black holes that may have preceded them. Collisions and mergers made the galaxies we see around us today. The rate of expansion seems to be accelerating, compliments of an even more mysterious "dark energy" about which we know little but its existence. What caused the Big Bang is unknown, and perhaps may never be, science always at a balance between knowledge and mystery.

Continuing on

But now back now to the main tour. As the Big Dipper is to northern hemisphere spring and summer (the Little Dipper for northerners

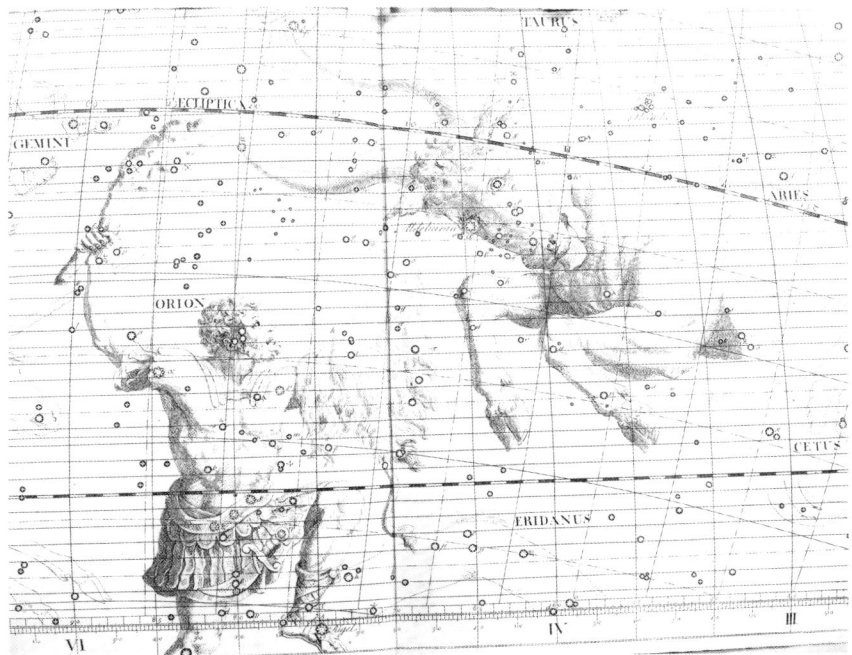

Figure 4.3. Orion, the Hunter, prepares to whack the Zodiac's charging Taurus, the Bull, in John Flamsteed's "Atlas Coelestis" of 1789. The drawings overlay highly accurate positionings of the stars and are meant to show the locations of the constellations. The red supergiant Betelgeuse marks Orion's right shoulder, while the blue supergiant Rigel shines at his left foot. In between is the three-star Belt, from which dangles the stellar trio that makes his Sword. The river Eridanus flows at bottom right, ending out of sight at bright Achernar. Rare Book and Manuscript Library, Univ. of Illinois at Urbana-Champaign.

circumpolar and always visible), Orion (the Hunter) is to winter, his three-star Belt of blue-white (and thus hot) stars prominently crossing the sky close to the celestial equator. He is said to have been poisoned by a scorpion that became the constellation Scorpius. The gods put them in the sky opposite each other so that Orion need not look upon his killer. In another story, he was struck down by an arrow accidentally shot by his lover, Diana. He's most famously depicted as raising his club against charging Taurus, the Bull, which lies to the northwest of him, and which is most promi-nently marked by a vee-shaped head, an old cluster of stars (the group born together) called the Hyades. Look down from the Belt to the three-star Sword he carries below. Even binoculars will reveal the center of the Sword to be surrounded by the Orion Nebula, a huge cloud of glowing interstellar gas lit by hot stars and that marks a great star-forming engine some 1500 light years away. At his right shoulder (he is facing you) is the zeroth magnitude star Betelgeuse. One of the larger stars in our galaxy, this cool red supergiant would nearly fill the orbit of Jupiter ("supergiant" a technical term implying great size and luminosity; see Chapters 6 and 7). At a distance of almost 600 light years, the great star shines with the light (including infrared) of 85,000 Suns, implying from theory a mass nearly 20 times that of the Sun. Elsewhere we see stars even larger and brighter. Across from the Belt, at the lower right corner, we find a bright blue-white supergiant, zeroth magni-tude Rigel, making the figure one of four that have two stars of first magnitude or brighter.

Down and to the left of Orion lies an opposite extreme. Here, in Canis Major, Orion's Larger Hunting Dog, shines Sirius, the bright-est star in the sky. It's not only intrinsically luminous (26 times the solar luminosity including ultraviolet), but it's close too, only nine light years away, making it nearly minus-second magnitude. Closely circling around it every half century is a tiny blip of a hot star 10,000 times fainter than Sirius itself, which requires it to be somewhat smaller than Earth. But from Kepler's laws and an assessment of the center of mass, we find a mass for the faint companion of about the same as that of the Sun. The average smeared out density of the Sun

Figure 4.4. The core of the Orion Nebula, which surrounds the Sword's middle star, is easily visible in binoculars. The gaseous nebula is the remnant of the active star formation that made the quartet of stars at the Nebula's center only a million years ago. The brightest of the four, Theta-1 Orionis C, powers practically the whole structure, which extends to more than a dozen light years out from the center and far off the page. About 1500 light years away, the nebula is a blister on the surface of the optically-invisible Orion Molecular Cloud, which pretty much fills the constellation. NASA/ESA, M. Robberto, (STScI/ESA) and the HST Orion Treasury Project Team.

is a bit more than that of water, which by definition is a gram per cubic centimeter (about the size of a sugar cube). Stuff all that into a body a hundredth the solar diameter and you get an average density of a ton per cubic centimeter. Sirius B, a "white dwarf," is the end product of stars like the Sun. Betelgeuse and Rigel, fine examples of dying massive stars that are done with core hydrogen fusion, are however destined to explode. Maybe tonight, maybe millions of years from now. They should produce remnants far smaller, neutron stars not much bigger than Manhattan, again revealing something of the immense range in stellar properties.

At the southwest apex of a bright triangle below Sirius you will find Adhara (Arabic for "the Virgins," which also applies to the whole triangle). Adhara (Epsilon Canis Majoris, Greek letters explained below) is the faintest of the first magnitude stars, making Canis Major another of the fab four with two of them. If you had ultraviolet eyes, this blue-white hot gem would be the brightest star in the sky. Betelgeuse and Sirius mark the northwestern and southern apices of the Winter Triangle (so-called by provincial northerners), which also includes bright Procyon in Canis Minor (the Smaller Dog) at the northeastern corner. Oddly, Procyon has a white dwarf companion too.

Figure 4.5.   The Summer Triangle, made of the bright stars Deneb (upper right), Vega (lower right), and Altair (lower left), descends the western sky. Deneb is the luminary and tail of Cygnus, the Swan, which flies down the Milky Way, the configuration ending at Albireo, the Swan's head. Down and a bit to the left of Deneb is Sadr (Gamma Cygni), from which stretch the bird's wings. Cygnus upside-down becomes the Northern Cross. To the left of Vega is the delightful parallelogram that makes most of Lyra, the Harp. The double-double star Epsilon Lyrae is just above Vega. The two stars flanking Altair make a pair of wings for Aquila, the Eagle, though classically they flap much farther out. Up and to the right of Altair is the small figure of Sagitta, the Arrow, while up and a bit left lies the exquisite set of five or six stars that makes Delphinus, the Dolphin. The Great Rift, formed of dark and cold interstellar clouds, runs down the Milky Way's spine. J. B. Kaler.

The Winter Triangle's warm-weather partner is the Summer Triangle, made of three white stars: zeroth magnitude Vega (in Lyra, the Lyre) at the northwestern apex, first magnitude Altair in Aquila (the Eagle) at the southern, and the first magnitude (though near the end of the list) supergiant Deneb (the tail of Cygnus, the Swan) at the northeastern, the triangle framing the Milky Way. Vega is surrounded by an infrared-radiating disk that implies a circulating planetary system, though no planet has ever been seen. Just northeast of Vega is Epsilon Lyrae, the original "double-double" star. To a sharp eye, Epsilon appears as a wide pair, while through the telescope each is closely double. The stars of each pair are in mutual orbits that take around a thousand years to complete, while the pairs themselves orbit each other over more than half a million. Roughly half of our galaxy's stars are in double or multiple systems. Application of Kepler's laws to orbiting pairs allows the measure of stellar masses. The first known real double was Mizar, the second star in from the end of the Big Dipper's handle, which is a backyard telescopic favorite. Each of its components has been found to be again double. Together with nearby Alcor, which is also double, the whole system is sextuple (Mizar and Alcor making the Arab's Horse and Rider). Lyra's main figure is a small delightful parallelogram. In it, between Beta and Gamma Lyrae (the two stars at the southern end of the parallelogram), is the Ring Nebula in Lyra, a tiny gaseous donut called a "planetary nebula" (having nothing to do with planets) that marks the death of a star (see Chapter 7). The western one of the two, Beta Lyrae, is an eclipsing double in which each member of the pair gets in front of the other with a period of 12.9 days, the bigger of the two eclipses changing the apparent combined brightness by nearly a full magnitude.

Deneb is seemingly fainter than Vega (25 light years away) only because of its great distance of 1400 light years. If you tip Cygnus upside down you get the Northern Cross, an asterism with Deneb now at the top. At the bottom of the Cross (the head of Cygnus) is pretty Albireo, a lovely example of a colored double star, one orange the other blue. A favorite while showing the sky to students or neighbors, if conditions are poor "there's always Albireo."

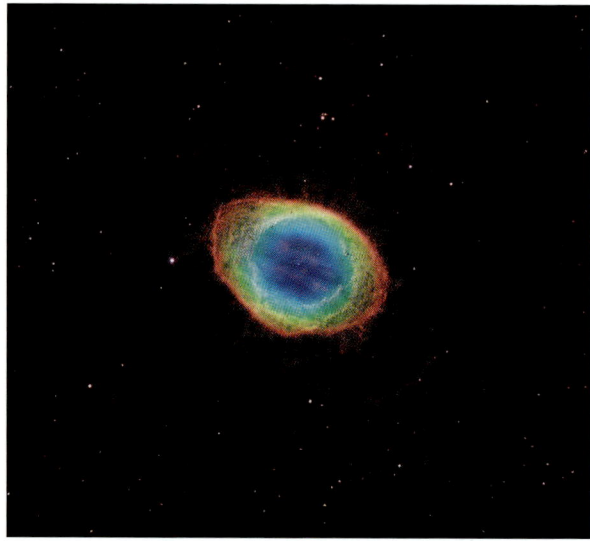

Figure 4.6.   The Ring Nebula in Lyra lies smack between the lower pair of stars in Vega's parallelogram (Figure 4.5). The bright ring, easily seen in a small telescope, was ejected from the faint star in the middle when it was a distended giant. The remnant core is hot enough to energize the ring out to half a light year or so. Surrounding the nebula are complex shells of gas from previous mass-loss episodes. The Sun may create something similar when it dies. NASA/ESA, C.R.O'Dell (Vanderbilt U.), and B.Thompson (LBTO).

If there be "supergiants," there must be smaller "giants." A prime example is Arcturus in Boötes, the Herdsman, who drives the Greater Bear around the pole. Just barely outshining Vega, orange Arcturus, the brightest star of the northern hemisphere, can quickly be found by following the curve of the Big Dipper's handle southward. And where there be giants there must be yet smaller "normal" stars, that like the Sun, Vega, Altair, and Sirius are quiet hydrogen fusers and that for historical reasons are oddly called "dwarfs." And there you have the basic kinds of stars. We'll look at the origins of them later. But in quick summary, stars are born as hydrogen fusing dwarfs. Lower mass ones, those under 8–10 times the solar mass, become first giants and then tiny white dwarfs. More-massive dwarfs (a spectacular misnomer) turn into supergiants like Betelgeuse, Rigel, and Deneb, and then explode as supernovae (and yes, there are lesser exploders, ordinary novae). Fun to watch, you would not want to be too close to a supernova.

Star names

Arcturus, Antares, Deneb, Zubenelgenubi, Epsilon Lyrae. Where do these peculiar star names come from? Some derive from the Greek words for the star's character or position. Sirius, the name of the brightest star in the sky, comes from the Greek for "searing," while Arcturus means "the Bear Watcher" ("arktos" is Greek for "bear"), the bright orange star following the beast around the pole. Reddish Antares ("ant-Ares," for the Greek god of war) fools us into thinking it's Mars. Admiring, indeed sheltering, Greek astronomy, the ancient Arabs added more names that applied to the Greek constellations and yet others that came from their own indigenous patterns. The Arabic names were later translated into Latin, often so badly and obscurely that major scholarship and a facility for several languages are needed to figure them out. The large majority of star names are thus of at least Arabic origin, even if now unrecognizable. "Betelgeuse" comes from "bet al Jauza," meaning "the hand of the central one" (whose identity is not known). More straightforward is "Deneb," Arabic for "tail," which is applied to the end of Cygnus, the Swan, and in one form or another to nearly a dozen other stars such as "Denebola" at the tail of Leo, the Lion, and Deneb Kaitos in the Whale's (Cetus's) tail. Everyone's favorite (and if not, it should be), Zubenelgenubi, is Arabic for the "southern claw" of Scorpius (the Scorpion), which actually partly makes up the constellation to the west, Libra, the Scales. (Zubeneschamali is the northern claw.)

This kind of naming gets pretty complicated. Plus it's hard to remember more than a thousand proper names. Though they are still used for the brightest stars and a few with special characteristics, something more logical was needed. In 1603 Johannes Bayer published his magnificent star atlas, the "Uranometria," in which he used Tycho's measures of position. The constellations are illuminated by superb wood block engravings according to their meanings. Within each constellation, he assigned Greek letters to the stars

more or less in order of brightness that go with the Latin posses-
sives of the constellation names, all of which now have three-letter
abbreviations. When Bayer ran out of Greek letters, he switched to
Roman. The brightest star in Canis Major, Sirius, then becomes
Alpha of Canis Major, or Alpha Canis Majoris (Latin possession
implied by case endings), which by internationally accepted abbreviations becomes Alpha CMa. The brightness rule, however, is
commonly breached and the letters allotted by position, as in they
are in the Big Dipper. Bayer also had ideas known only to himself.
Other astronomers later extended the scheme to the southern
hemisphere, which neither Bayer nor Tycho could see (though
Bayer tried to include the deep south via descriptions made by
early travelers).

Bayer's system, still in wide use, was followed about a century
later by one derived from the work of John Flamsteed (1646–1719),
the first Astronomer Royal of England. Following a naval disaster,
Flamsteed was given the task of telescopically measuring accurate
positions for a large number of stars to be used for navigational
purposes. His work was lifted by Newton and Halley, who published
it early and added numbers from west to east within the constellation of residence. (Flamsteed was furious and burned nearly all the
copies.) Vega, Alpha Lyrae ("Alpha of Lyra," the Lyre), is thus also
3 Lyrae. Shifting constellation boundaries, which were finally
secured around 1930, produced anomalies in both systems, leading
to "orphans" in which stars that carry one constellation name wind
up in another. Beyond these names are those from dozens of specialized catalogues, notably the Yale "Bright Star Catalogue," in
which some 9000 stars down through sixth magnitude are assigned
"HR" numbers from west to east starting at the Vernal Equinox,
which makes Arcturus HR 5340 ("HR" amusingly standing for
"Harvard Revised"). The Henry Draper spectrographic catalog of
the early twentieth century (Chapter 6) gives widely-used "HD"
numbers for 350,000 stars to roughly the tenth magnitude (Arcturus
also HD 124897). Adding up all the catalogs, brighter stars can have
as many as 50 names.

The Zodiac, astrology, and black holes

Go down the shining path of the Milky Way to find the scary figure of Scorpius, the Scorpion (within which resides the bright reddish supergiant Antares), Orion's nemesis. Looking just like what it is supposed to be, Scorpius is one of the constellations of the Zodiac. These, like Leo (the Lion), Aries (the Ram), Taurus (the Bull), etc., embrace the ecliptic and are the foundations of the ancient pseudo-science of astrology. In western constellation lore, there are traditionally 12 of them, the rounded-off number of times the Moon goes through its phases (Chapter 2). Holding the moving Sun, Moon, and planets (all perceived as gods), the zodiacal figures not only tell the old stories, but are also sacred. All but Libra are animalistic, "Zodiac" having the same root as "zoo" and "zoology" and meaning "circle of animals." As augured above, even Libra falls into the set, as with Zubenelgenubi and Zubeneschamali it was once (and for that matter still is) the Scorpion's claws.

Simple Sun-sign astrology (what you find in newspapers and magazines, not to mention websites), wherein your daily, even eternal fate is determined by the zodiacal location of the Sun at the time of your birth, is but the tip of the pseudo-iceberg. ("Astrology took a major step toward achieving credibility today when, as predicted, everyone born under the sign of Scorpio was run over by an egg truck": from a cartoon in "Sky and Telescope," August, 1989.) To the believer, the planets, named after Roman versions of Greek gods, have specific influences upon us that are modified by their positions relative to one another and to the Sun and Moon, which also represent celestial deities. Their power is sent to the individual through 12 fixed "houses" that divide the celestial equator, one for love, another for money, and so on. The astrologer calculates where all these are positioned at the "native's" (the "mark's"?) birthday and gives out various pronouncements about his or her personality and fate. There is far more to it that includes the four ancients elements of earth, air, fire, and water (never mind the actual 92 or so, but we digress...) How the planets actually DO this seems to be left largely unsaid.

Figure 4.7. Looking remarkably what it is supposed to be, Scorpius in mythology killed Orion. Set into a beautiful part of the Milky Way, Scorpius is filled with hot blue massive stars. With the red supergiant Antares at its heart, the creature's Stinger is represented by the pair of stars toward lower left. To the left of the Stinger is the bright "open" star cluster Messier 7, and above that, another open cluster called Messier 6. Several dark dust clouds, cold cauldrons of star formation, block the light of distant stars. Note particularly the long "Pipe Nebula" near the left center edge. J. B. Kaler.

It's vital to distinguish the astrological "signs" of the Zodiac from the actual constellations that make the astronomical Zodiac. The signs are uniform 30-degree-wide segments of the ecliptic that once more or less overlaid, and took their names from, the astronomical constellations. Compliments of precession (the 26,000 wobble of the

Earth's rotational axis around the orbital axis, discovered by none other than Hipparchus more than 2000 years ago), they no longer do, but are shifted one constellation over, the sign of Scorpius now sadly lying on its own astronomical claws in Libra. Libra is Libra because it once held the Autumnal Equinox ("balance of days" and all that), which is now one constellation to the west, in Virgo. Amusingly, the press discovered the effect of precession early in this century and made a bit of a fuss over it, without of course mentioning Hipparchus. That little matter of course is not why astrology doesn't work.

Look around the neighborhood. To the north of Scorpius is the sprawling figure of Ophiuchus, who is wrapped by Serpens, the Serpent, the only constellation to come in two disconnected parts, separated by the Serpent Bearer: the Head (Serpens Caput to the west) and the Tail (Serpens Cauda to the east). Bayer and Flamsteed treated it as one constellation and the names and numbers flip back and forth between Caput and Cauda. Ophiuchus is loosely related to the Lacoön, a figure of the Trojan War whose statue resides in the Vatican. The Lacoön descends to us through the caduceus, the snake-wrapped staff of the physician, and the "Spirit of Communication," a heroic figure entwined with a thick cable who once represented the Bell Telephone Company. With his modern boundaries lying across the ecliptic, Ophiuchus is sometimes called the thirteenth constellation of the Zodiac, though he really does not belong there. Nevertheless, there are people who consider themselves "Ophiuchans" and who get quite upset if you dismiss him.

East of Scorpius, within the heart of the Milky Way and within the Zodiac, lies Sagittarius, the centaur Archer, whose outline is best known for its charming "Teapot" and upside-down "Little Milk Dipper." Sagittarius contains a treasury of bright interstellar clouds, or nebulae, of which the Lagoon and Trifid Nebulae (Chapter 5) are the most prominent. Sagittarius's greatest possession is the center of the galaxy (rather, the direction to the center), first seen as a radio source called "Sagittarius A." Later radio observations with continental-sized interferometers found much of the radiation to come from a near-point-source called Sagittarius A*. At a distance of 26,000 light years, the galactic center is impossible to see in the optical

Figure 4.8. The brightest part of the Milky Way runs through Sagittarius (the Archer). The constellation is best known for its bright upside-down "Little Milk Dipper," which lies up and to the left of center. Down and to the right of center, dust clouds obscure the center of the galaxy, which contains a black hole of four million solar masses 25,000 light years away. The bright red patch up and to the right of center is the Lagoon Nebula, Messier 8; the Trifid Nebula (Messier 20) is just up and to the right of it. Both harbor dark clouds in which stars were recently (and are still being) born. The open cluster Messier 7 in Scorpius is just below center, while the scorpion's Stinger falls near the lower edge. The globular cluster M 22 is the bright "star" just up and to the left of the Dipper's handle. J. B. Kaler.

spectrum due to the dimming effects of interstellar dust along the line of sight (Chapter 5). However, infrared observations (which can penetrate the dust) of stars in orbit around it show Sagittarius A* to weigh in at four million solar masses, all of it stuffed into a body much smaller than the orbit of Mercury.

Our only conclusion is that Sgr A* must contain a supermassive black hole, a body so dense that light cannot escape its gravitational grip. Toss a ball in the air. It comes back. But throw it up at 7 miles (11 kilometers) per second and, though it continually slows down, it will never stop and return. Squeeze the Earth, make it smaller, and (because you are closer to its atoms) the gravity at the surface increases, so you have to throw the ball faster to get it to escape. When the Earth is about the size of a golf ball, the escape speed hits that of light. No radiation can get out and the Earth becomes a black hole. It has all of its gravity: an outsider just could not see it. It's admittedly an unlikely scenario, but the evidence for the existence of real black holes of stellar sizes as well as for the "monsters of the midway" at the centers of galaxies (some of which dwarf our own) is overwhelming (Chapter 7).

Though the galaxy's central black hole is itself invisible, its surroundings are very much not. Gravitational interactions among closely orbiting stars may bring one too close, whereupon powerful tides shred the intruder and then cast it into a brilliant surrounding disk, from which hot gas pours out in jets flowing perpendicular to the disk as well as into the black hole, increasing its mass. The central massive black hole of a galaxy (rather, its environs) is among its most luminous features and can be so energetic as to influence the whole system. Indeed, its jets can even affect neighboring galaxies. Compared to others, our central black hole is small, some going into the billions of solar masses. Stellar black holes are known through their membership in double star systems, in which the black hole tidally distorts and sucks mass from a dying companion, usually a supergiant, which again lights up the surroundings.

By coincidence, the Winter Solstice is pretty closely aligned in the sky with the center of the galaxy. Because of precession of the Earth's axis, such a solsticial alignment happens every 13,000 years. This particular passage, which more or less coincides with the ending of the Mayan "long count" calendar, signified to the less aware the end of the world. Among other things, in 2012 it was supposed to have caused the Earth's rotation, or magnetic field, or both (it's unclear) to reverse, buildings to fall, continents to split, and the world to be destroyed. Apparently, this happens periodically. There is no ancient evidence for such a happening, it's physically impossible, and if still in doubt note that we are still here. (And yes, astronauts did land on the Moon, there is no vast conspiracy that covers up that it was all done in Arizona, etc.)

Striding the sacred path

Follow the Sun. From Sagittarius we pass eastward into Capricornus (the unlikely Water Goat), which a few thousand years ago chewed at the Winter Solstice. Hence the Tropic of Capricorn, the circle of latitude 23.4 degrees south of the equator that marks the southern limit of the overhead Sun. If named today, the circle would be the Tropic of Sagittarius. Climbing northeasterly, we enter Aquarius, the Waterman, then Pisces, the sprawling Fishes (which lie south of Pegasus and Andromeda, who appear in a later scene) who nibble at the Vernal Equinox. A couple thousand years ago, the equinox was in Aries, the Ram, a powerful fertility symbol that tops off the newspaper astrology columns since they are a couple millennia out of date. Some 2000 years hence, precession will bring the equinox into Aquarius, which will presumably begin an era of peace and love, as expressed by "The Age of Aquarius" from the first rock musical, "Hair." One hopes sooner than that.

Aquarius is usually depicted as pouring water from his jug (a prominent "Y"-shaped asterism) into the mouth of Piscis Austrinus, the Southern Fish. Together, Capricornus, Aquarius, Piscis, and Piscis Austrinus make the "wet quarter," an allusion to an ancient rainy season. Piscis Austrinus is best known for the first magnitude star Fomalhaut, from Arabic for "the Fishes Mouth," about which orbits a visible planet (most planets detected indirectly, as seen in Chapter 8).

We are now well into northern celestial territory where east of Aries we again encounter Taurus, the Bull, a magnificent constellation that looks its part (look back to Figure 4.3). Taurus is especially known for the first magnitude giant Aldebaran ("the Follower") and two prominent ragged "open clusters," the Pleiades (the Seven Sisters, who Aldebaran follows) and the Hyades, which makes the Bull's vee-shaped charging head. In mythology, the Pleiades were the daughters of the god Atlas and mortal Pleione (who are represented by name); the Hyades, also fathered by Atlas, are their half-sisters. When seen in northern autumn evenings, they are all harbingers of cool days to come. To the natives of South America they were a harvest basket.

Figure 4.9. Among the most beloved sights the sky offers is the Pleiades (Seven Sisters) cluster in Taurus, made of hot blue stars (and many fainter cooler ones) just 400 light years away. Only 130 million years old, the cluster is almost new-born, at least as compared to the 4.5-billion-year-old Sun. Sprawled across a couple degrees of sky, it's a wonderful sight in binoculars. The cluster is passing through a dusty cloud whose tiny grains reflect the blue starlight. Mark Killion.

Open clusters, which contain from a few to a few thousand stars, flock the Milky Way. Most stars, including the Sun, seem to have been born as members of clusters that eventually fall apart. Long photographic exposures of the Pleiades reveal delicate wisps of interstellar dust that reflect bluish starlight from hot stellar members, the cluster on its way through a dusty interstellar cloud. The Hyades, just 150 light years away, is one of the closest clusters, while the Pleiades lie at almost thrice that distance. Aldebaran, which makes the Bull's eye, appears smack in the middle of the Hyades, but only by coincidence. Roughly halfway to the cluster, it's not part of the assembly.

Open clusters are distinctly different from the other kind, the rare great "globular clusters" that jam hundreds of thousands, even millions, of stars into much the same space. While open clusters are denizens of our galaxy's star-forming disk, the globulars

Figure 4.10. The globular cluster Messier 80 in Scorpius is one of 180 or so ancient systems that include Messier 22 in Sagittarius (see Figure 4.8), M 30 in Capricornus, M 3 in Canes Venatici, the deep southern hemisphere's 47 Tucanae, and the Great Cluster in Hercules, M 13. With only a few tens of thousands of stars, M 80 is one of the smaller classical globulars. In the largest, nearly a million stars can be packed into a stellar cloud only a few tens of light years across. ESO/M.-R. Cioni/VISTA Magellanic Cloud Survey.

inhabit a vast surrounding galactic halo. They are so bright that they can easily be seen over vast distances. The best known in the north is the Great Cluster in Hercules, a somewhat dim ancient constellation to the northeast of Arcturus. The cluster, 25,000 light years away, is better known as Messier 13, after Charles Messier (1730–1817), who compiled a catalogue of about a hundred celestial objects that might be confused with comets and that include nebulous clouds, other galaxies, and brighter clusters. (The Orion Nebula is Messier 42, the Pleiades M 45, and the Ring Nebula M 57.) The brightest globular cluster is Omega Centauri, in the southern hemisphere's Centaurus, the Centaur, the cluster easily visible to the naked eye. Both Sagittarius and Capricornus have really nice ones, M 22 and M 30. Seen against the background of

Figure 4.11.   The Crab Nebula in Taurus is the remnant of the great supernova, the Chinese "Guest Star," of the year 1054. Expanding at 1500 kilometers per second, 6500 light years away, it's now almost 10 light years across. The nebula is powered by a neutron star at the center that is only 30 kilometers across and spins (and flashes) 30 times per second. The Crab not only carries a load of heavy elements made in the explosion to mix with the dark clouds of interstellar space, its gigantic shock wave will help compress them to help make new stars. It also accelerates atomic particles to near the speed of light, creating the ubiquitous cosmic rays. Over the aeons, supernovae have caused the heavy-element content of the Milky Way to increase with time. NASA/ESA, J. Hester and A. Loll (Arizona State U.)

the Milky Way, M 22 is a stunning sight. The closest open cluster is made of the five middle stars of the Big Dipper along with numerous fainter ones.

Taurus also embraces the amazing Crab Nebula, Messier 1. It's the rapidly expanding gaseous remnant of a massive star that exploded as a supernova in the year 1054. In spite of it being 6500 light years away, at its peak the star shone with the light of Venus (the event well recorded by the Chinese and others). Expanding at

1500 kilometers per second, the cloud is now some 10 light years across. (Remember that we ignore light travel time; things happen as we see them.) At the Crab's center is a collapsed neutron star with an average density of a million metric tons per cubic centimeter, a million times that of Sirius's white dwarf. About 20 kilometers across, containing more than a solar mass, with a magnetic field a trillion times that of Earth, the neutron star spins 30 times per second and powers the whole nebula. Radiation beamed along the tilted magnetic field axis makes it appear to flash on and off at the same rate, rendering it a "pulsar," one of more than 1000 known. A bit more massive and the Crab Pulsar would fall into its own black hole. Neutron stars are the collapsed old nuclear-burning cores of stars born with at least 8–10 solar masses, of supergiants like Betelgeuse and Deneb, which are destined to explode as supernovae. Under the right circumstances (Chapter 7), white dwarfs can go off too. Nearby Sirius B and Procyon B are not among this rarefied group, so don't worry.

Then it's over the top with Gemini, the Twins, one a god, one not (in Greek mythology it somehow makes sense). The constellation is highlighted by the first magnitude orange giant Pollux (the god) and the brightest star of second magnitude, Castor (not). Pollux has a massive planet (known by the star's motion), while Castor is another sextuple. Traditionally, Gemini is noted for the Summer Solstice, which according to modern boundaries just recently moved into Taurus (an artifice we'll ignore). In the ancient past the solstice was in the next constellation over, Cancer (the Crab, having nothing to do with the Crab Nebula), hence the northern latitudinal limit of the overhead Sun, the Tropic of Cancer. (The constellation also has nothing to do with the disease, though you should either stay out of tropical sunlight or use sunscreen, otherwise you are subject to skin cancer. But it was such a scary sign that some astrologers started to call Cancer people "Moon Children." It didn't catch on.) Cancer is also noted for its prominent central star cluster, the Praesepe (an allusion to the Christian manger) or Beehive (M 44). Close to 600 light years away, it's a delightful sight in binoculars.

Leo, with bright Regulus (a rare Latinate name here meaning "the Little King"), next roars. The constellation represents the Nemean lion slain by (here he is again) Hercules. Regulus too has a white dwarf companion, showing how common the little things are. Submassive, the white dwarf may once have passed some of its matter onto Regulus proper. At the Lion's tail is second magnitude Denebola, which, like Vega, features some kind of debris disk, suggesting planets. Then it's down through Virgo, the estimable Virgin, known principally for Spica (among the hottest of first magnitude stars, and a close double as well), its huge and relatively nearby (45 million light years away) cluster of galaxies, and the Autumnal Equinox, which it got from Libra.

Andromeda and family

The last supernovae to be seen in our galaxy were Tycho's Star of 1572 in Cassiopeia and Kepler's Star of 1604 in Ophiuchus. Cassiopeia brings us to a whole bunch of related figures, those of the Andromeda myth, all nicely on display in northern autumn skies. Queen of Ethiopia, wife to King Cepheus, Cassiopeia irritated Neptune by bragging that her daughter Andromeda was lovelier than the sea nymphs. You don't mess with the gods. As punishment, Neptune ordered the maiden to be chained at the coast to be devoured by Cetus, the Whale or Sea Monster. But along came Perseus on Pegasus (his Flying Horse), who was returning home from slaying the dreaded Medusa. With hair of snakes, one look at her would turn you to stone. Perseus saw the young lady, and what's a hero to do but rescue her by showing Cetus the Medusa's head, which sight slew the monster, following which the couple happily wed.

Perseus, Cassiopeia, and Cepheus are in the Milky Way, the latter two the most northerly outposts along the starry stream. Perseus is best known for its unique double open cluster (not surprisingly called the Double Cluster) and for Algol, the Demon Star (the Medusa herself, but now safe to see), Beta Persei, in which mutually orbiting stars eclipse to cut the light in half every 2.9 days. The Double Cluster, 7700 light years away, is a jewel box filled with

Figure 4.12. The Andromeda Nebula, Messier 31, is another galaxy, one very much like ours with a disk and surrounding sparse halo. Just over two million light years away, it's the farthest thing you can easily see with the naked eye. It has two bright companions, the small spherical galaxy M 32 at the lower right edge of M 31, and the small elliptical system near the top. The outer parts of M 31 are blue as a result of current star formation that includes massive blue, hot stars, while the inner portions are reddish from the light of older stars. M 31 is a classic sight for the amateur telescope or even binoculars. It's expected to collide with us in a billion years or so. Mark Killion.

contrasting red supergiants and blue dwarfs. Perseus is also famed for its spectacular collection of hot massive young stars. Andromeda contains the aptly named Andromeda Nebula (M 31), an easily visible spiral galaxy like ours two million light years away that has already been introduced. Cetus is celebrated for Mira ("the Wonderful"), a long-period variable star (an actual category, LPV, not just a description) that is a bright part of its constellation for a month or two and then completely disappears from view in a nearly year-long cycle. It's in the act of dying. Pegasus contains the first star with a known orbiting planet (51 Peg) and a beautiful compact globular cluster, M 15. (The mythical animal has been passed down to us as the symbol of an oil company.) Among Cepheus's possessions is a variable star, Delta Cephei, which holds the key to the distances to far-off galaxies. The most prominent member of the class of Cepheids, Delta Cep goes between fourth and fifth magnitude over a period of 5 days. The variation period of a Cepheid gives its absolute stellar magnitude, which with apparent magnitude yields distance. Cepheids in the Andromeda Galaxy were used around 1925 by Edwin Hubble to make the first measure of another galaxy's distance and to prove that it and countless other fuzzy blobs are really vast systems external to ours. Delta Cep and the rest of its kind are indispensable to modern cosmology. Polaris is another, though minimal, example. Cepheus also has two of the largest red supergiants (and thus stars) known, Herschel's Garnet Star (Mu Cephei) and VV Cephei, another eclipser whose bigger member approaches the size of Saturn's orbit.

While in between

Just south of Andromeda, between it and Aries, lies a trio of stars that just begged to be called Triangulum. It's notable for another galaxy that under fine conditions is also visible to the naked eye. The Triangulum spiral, Messier 33, is roughly the same distance as the Andromeda Galaxy and, though much smaller than M 31 or ours, is a significant part of the Local Group.

Other constellations surround those highlighted above, some big, others tiny. Between the Bears winds lengthy Draco the Dragon, while on a line from Cygnus to first magnitude Altair (flanked by two stars that make the trio look like a flying bird; today it might be called "Airplanus" or some such) lie little Delphinus, the Dolphin, which looks more like a hand with a finger pointing south, and Sagitta, the eponymous Arrow (see Figure 4.5). South of Cancer, Leo, and Virgo slithers the longest constellation of the heavens, Hydra, the Water Serpent, which, running precession backwards, thousands of years ago basked along the celestial equator. Its head appears to look back at you with malevolent eyes. On its back are Crater, the Cup, and to the east the distorted box that makes Corvus, the Crow, or Raven. Crater vies for dimmest ancient constellation with Equuleus, the Little Horse, southeast of Pegasus. The top stars of Corvus point eastward back to Spica. To the Navaho, Corvus was known as "man with feet ajar," showing that different folks can have wildly different views.

To the southeast of Corvus, below Virgo and Hydra's tail, is huge Centaurus, the Centaur, which for northerners plunges far below the southern horizon. It's another of the four constellations to be marked by two first (or brighter) magnitude stars: nearby Alpha, whose proper name is Rigil Kentaurus (the "foot of the Centaur"), and Beta Cen. Centaurus also has the brightest globular cluster, prominent Omega Cen, one of the few non-stellar objects to carry a Greek letter. It's now widely believed not to be a native cluster but to be the core of small galaxy that long ago merged with ours, showing how complicated stellar configurations can get. Between it and Scorpius howls Lupus, the Wolf, the three constellations especially known for their collections of hot, massive, young, blue stars, many of which are destined to explode as supernovae.

On the other side of the southern sky, south and southwest of Canis Major (and Sirius), sails vast Argo, the Ship of the Argonauts. It's so big it's now separated into three still-large constellations, Vela (the Sails), Puppis (the Stern), and Carina (the Hull), which has the sky's second brightest star, minus-first magnitude Canopus, in its hold. Like Centaurus, Scorpius, and Lupus, the three parts are filled with massive blue stars that include one of the heaviest known. Around 1845, Eta Carinae, a massive double star in mutual orbit, burst from fourth magnitude to become the second brightest star of the sky, nearly rivalling Sirius. By the end of the century, in part as a result of a dust cloud of its own making, it had faded to invisibility and is now slowly recovering. As a constant theme, all stars lose mass as their fuel runs out and then first expand and then die. In its outburst, which is not understood but may be related to its duplicity, Eta must have burped a solar mass or more that is now seen expanding about the supergiant within. It's buried within the vast Carina Nebula, which highlights a bright portion of the southern Milky Way. As they say, "watch this space," as something is going to happen.

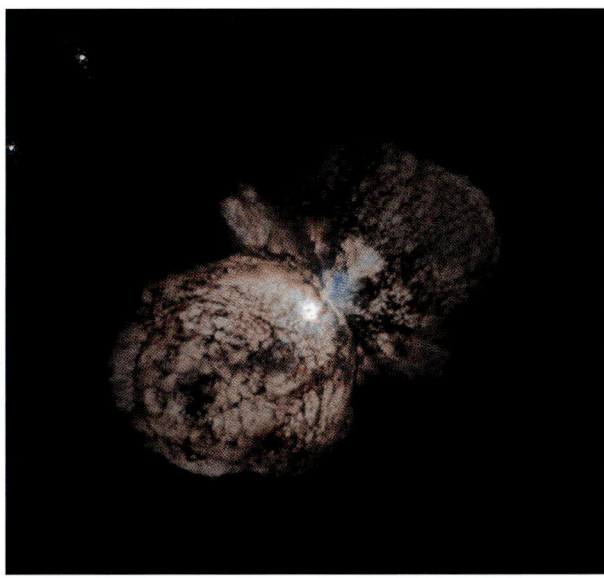

Figure 4.13. Six thousand light years away and among the most massive stars of our galaxy, Eta Carinae (in Carina, the Keel of the Ship Argo) carries about 100 times the mass of the Sun and has a companion with not a lot less. In the nineteenth century, Eta Car erupted, belching the bipolar dusty nebula that surrounds it. It may produce a "hypernova" and gamma ray burst concentrated along the rotation axis as one of the inner stars collapses into a black hole (Chapter 7). Six thousand light years away, Eta Car could pose a danger to Earth. Fortunately, its shattering beams will most likely be pointed elsewhere. Jon Morse (U. of Colorado) and NASA.

Among the other ancient delights is a pair of semicircular crowns in opposing hemispheres. The one in the north, Corona Borealis east of Bootes, honors Ariadne of the "Theseus and the Minotaur" story, who was summarily dumped by the hero. The other, Corona Australis, is just south of Sagittarius and is purported to be the Archer's crown. Corona Borealis has a weird sixth magnitude variable star, R CrB, which unpredictably puffs out a dust cloud that makes the star dim by a factor of more than a thousand.

The moderns

The modern constellations introduced in the seventeenth and eighteenth centuries are, to say the least, a mix. The northern ones (which include that part of the southern hemisphere seen from classical lands) mostly feature stars faint enough that they were ignored by the ancients, falling into their "amorphotoi," the unformed. Presumably they were part of a canvas upon which the gods could paint more stories and honor new heroes. Examples of the "new" (it's all relative) figures are Camelopardalis (the Giraffe), the obvious Lynx, Monoceros (the Unicorn, which actually harkens back to older times), Fornax (the Furnace), Microscopium and Telescopium, and so on. They reflect the interests, technologies, and life of the times. Not to mention attempts to honor or even gain the favor (and perhaps coin) of one's rulers. Within the group are Canes Venatici (the Hunting Dogs), which features a pair of stars (trio if you like) just south of the Big Dipper's handle that honors England's Charles II, and Scutum (the Shield of John Sobieski (1629–1696), the heroic King of Poland who fought off the Turks at the battle of Vienna in 1683). The latter is in a wonderful part of the Milky Way south of Aquila and north of Sagittarius. Then there was Halley's "Charles' Oak," which like the tree itself is among the fallen.

A little farther south of Canes Venatici is an especially fine modern that actually has a long heritage back to classical times but was not included by Ptolemy. Coma Berenices is a lacy group of stars that represents the hair of the ancient Queen Berenice, and is yet another naked eye open cluster 280 light years away. Too big to admire with a telescope, it's a wondrous sight as seen through binoculars. Like the Hyades, it's too large in extent to have been included in the Messier catalogue.

Many are the discarded figures, those that did not succeed and that include such gems as Officina Typographica (the Printing Office), Bufo (the Toad), the unfortunate Musca Borealis (the Northern Fly), and several nationalistic figures. One that almost made the "received list" comes from old Roman times. Antinous, who served Hadrian and is said to have committed suicide in the belief that his remaining years would go to the emperor, was constructed from the stars of southern Aquila and in Bayer's great atlas is seen carried by the giant Eagle. Antinous was eventually dropped by the bird. Dozens of others are of historical interest only, the International Astronomical Union accepting just 38 of the moderns in the 1930s.

The deep southern hemisphere is different, as there are several bright (as well as faint) patterns that could not be seen and recorded by the ancients. The southern moderns include animalistic figures, some like Phoenix fanciful, others like Grus, the Crane, quite real, as well (why, oh why?) Musca, the Fly. In reflection of Musca Borealis in the north, it was originally Musca Australis, but it now flits solo. Equally important are a variety of artifacts such as the imaginative Triangulum Australe (the Southern Triangle), Octans the Octant, within which we find the South Celestial Pole, Pyxis (the Compasses), which is linked with Argo, and so on. Octans joins Sextans the Sextant near Leo, and Quadrans, the Quadrant, which lies off the handle of the Big Dipper. The first two made it to the list of 88, but having enough of navigational instruments, Quadrans walked the plank. It still appears, however, as the apparent origin of January's Quadrantid meteor shower.

Of special interest are Dorado (the Swordfish) and Tucana (the Toucan), which respectively enfold the Large and Small Magellanic Clouds, small irregular satellite galaxies of our own large system. Averaging about 180,000 light years from Earth, both are easily visible to the naked eye, and with the Andromeda and Triangulum spirals bring the non-telescopic count up to four. Within the Large Cloud we find a massive interstellar nebula, 30 Doradus, that dwarfs the Orion Nebula (the number not from Flamsteed's work, as he could not have seen it). Close by, Supernova 1987a erupted to third magnitude in (as expected) 1987, becoming the first such explosion to be visible to the naked eye since Kepler's Star of 1604. Next to the Small Cloud, though in our galaxy and much closer, is the grand globular cluster 47 Tucanae, number two in glory after Omega Centauri.

The best known modern constellation is a hybrid, a configuration known since ancient times but re-worked from the feet of the

Centaur as Crux, the Southern Cross. At the southernmost reach of the Milky Way (opposing Cassiopeia and Cepheus in the north) and the last with two stars of first magnitude, Crux is a spectacular figure, especially when combined with first magnitude Alpha and Beta Centauri to the east of it, the configuration presenting an unforgettable sight. Tucked into it is a striking open cluster bright enough to have obtained a Greek letter from its brightest star, Kappa Crucis. Nearby in the Milky Way lies a black blob, the Coalsack, a "hotbed" (really a "coldbed") of star formation. It and other dark clouds in the southern Milky Way are so obvious that the Incas made "constellations" out of them, the Coalsack called "Yutu," a small bird.

Just four light years away, Alpha Cen is usually called the closest star to the Earth. But not quite. Orbiting the naked eye star, third brightest in the sky, is a dim 11th magnitude low mass dwarf (Chapter 6) companion to Alpha called Proxima Centauri that is by a few thousand Astronomical Units the actual closest star and that takes maybe a million years to orbit its bright companion. A closer look then shows Alpha itself to be double, the two taking 84 years to orbit each other. The brighter member is a near solar clone. A planet was once thought to orbit the somewhat dimmer member of the pair, but the discovery seems to have been spurious. The question yet lingers: Is anybody there? For that matter, is anybody anywhere? Maybe we can find out. So we now leave the tour and approach the stars themselves.

5

Out of Darkness

Following the credit roll of "Chicken Run," an animated film based on "The Great Escape" (and after most of the audience has escaped as well), the chickens debate the philosophical, indeed biological, question: "Which came first, the chicken or the egg?" Transmogrifying the barnyard into the galaxy: Which now do we look at first, stars or the eggs that hatch them? Stars are born out of the dusty gases of the interstellar medium (ISM), the term for all the matter that lies in the vast spaces between them. And vast it is; the distance to the nearest star is 30 million times the solar diameter. Yet as sparsely scattered as they are, as they die stars profoundly affect the interstellar medium and help drive star birth. The result is a continuous cycle, though with important dropouts to be taken up later. We have to break in somewhere, so let's pick the ISM and the creation of stars. In the next chapters, we'll look more closely at the kinds of stars that come out of it and how they live their lives and finally expire.

The interstellar medium is a stunningly complex mixture of hot, warm, cool, and cold gas. Like the stars, it's composed of 90 percent hydrogen and 10 percent helium with the usual tiny but overwhelmingly important mix of everything else, into which is churned a salting of "dust" made of tiny solid grains. Much smaller than the dust on top of your refrigerator, the grains are individually too small to be seen without a microscope. Hardly minor, the ISM holds roughly 15 percent of the visible mass of our galaxy, almost all of it in the flat disk. The average density of the gas is but one hydrogen

Figure 5.1. The vast Carina Nebula of the far southern Milky Way, more than 6000 light years from us, contains an extraordinary number of interconnected bright diffuse nebulae as well as numerous dark clouds that harbor star formation and that have birthed the massive hot stars whose ultraviolet light in turn illuminates the bright clouds. Some of the galaxy's most luminous and massive stars reside here, including Eta Carinae (Figure 4.13). NASA/ESA, N. Smith (U. of California, Berkeley), and the Hubble Heritage Team (STScI/Aura).

atom per cubic centimeter, making it a "vacuum" far better than can be achieved in any laboratory. But with a diameter of 100,000 light years and a disk thickness of a few hundred light years, with a billion billion centimeters per light year, the galaxy has a LOT of cubic centimeters, so it all adds up to a pretty significant mass.

Average, though, does not mean much, as the ISM is extraordinarily lumpy and varied, ranging from thick clouds full of molecules with densities up to a million times average to vast hot voids with almost nothing in them at all. About one percent of the mass of the ISM lies in the dust grains, most of which are based on either carbon or on oxygen and silicon, both kinds mixed with heavier elements, including calcium and real metals like iron, titanium, and nickel. Since the dust has condensed from the gas, the gas is depleted in these same heavy elements, though the average composition of the dust plus gas is similar to that of the Sun and stars, which makes sense since the stars have come out of it. On average the galaxy's disk contains about one dust grain per cubic meter, most of it coming from the winds of dying stars. That average does not mean much either. The dust is crucial to star birth, but is a serious pain to optical astronomers, as it dims distant stars, making them look too far away. At its worst, it totally blocks the view.

Cloudy skies

The tour of the sky in Chapter 4 visited a number of nebulae, Latin for "clouds." The term is, or at least was, loosely applied to any fuzzy celestial object and can be confusing as it includes many different kinds. The Andromeda Nebula (Messier 31: Figure 4.12) is another galaxy, as are M 33 in Triangulum, M 51 (the Whirlpool Nebula) in Canes Venatici, M 101 (the Pinwheel) and M 81, both in Ursa Major, and M 87 in Virgo, along with seemingly endless numbers of others. Under a dark sky, M 31 and much fainter M 33 can be seen with the naked eye. Those with spiral arms (which include the first four above) fall into an archaically-named subgroup of "spiral nebulae" whose natures were hotly debated until Edwin Hubble showed them in the late 1920s to be "extragalactic nebulae," that is, other galaxies quite distinct from our home galaxy. They have their own interstellar media, those in the spiral galaxies similar to ours.

Within our galaxy we earlier trolled past the Orion Nebula, M 42, which lies in the middle of the Hunter's Sword (Figure 4.4). It's an example of a diffuse nebula, a relatively dense gaseous cloud illuminated by one or more hot stars. About a degree across, the Orion Nebula is a spectacular sight in a small telescope, even in binoculars. Diffuse nebulae are among the most obvious and beautiful manifestations of the ISM. Bright ones stud the Milky Way (as they do in other spiral galaxies; Figure 4.2), most glowing with a reddish color, though some of the brightest, like the densest portion of the Orion Nebula, can take on a greenish cast. In Sagittarius we find the Lagoon and Trifid Nebulae (M 8 and M 20; Figure 4.8), in nearby Serpens the Eagle Nebula (M 16), in Cygnus near Deneb the fainter but larger appropriately-named North America Nebula, not far away in Perseus the California Nebula, in Monoceros near Orion the Rosette Nebula, deep in Argo the vast bright Carina Nebula, and hundreds of others with just catalogue names, thousands with no names at all. They range in size from tiny motes to huge structures a hundred or more light years across. The Orion Nebula, 1300 light years away, is around 25 light years in diameter (depending on how you define its limits), six times the distance between here and Alpha Centauri. If the Tarantula Nebula in the Large Magellanic Cloud (Chapter 4) were substituted for it, the nebula would fill the entire constellation. NGC 604 in Triangulum's spiral galaxy M 33 is similar.

The most obvious diffuse nebulae carry proper names that go back more than a century. Proper names for the fainter ones, some of which are pretty silly (but why not), continue to proliferate. More order is brought by various catalogues. The most basic is the short Messier list of 110 objects (see Chapter 4), which is a favorite among amateurs. Far more extensive is the "New General Catalogue" of nearly 8000 celestial objects that was compiled in the nineteenth century by the Danish-Irish astronomer J. L. E. Dreyer (1852–1926). It's an extension of John Herschel's earlier "General Catalogue of Nebulae and Clusters of Stars," which in turn was based on William and Caroline Herschel's (John's father and aunt: see Chapter 2) catalogue of discoveries. The "NGC" was later supplemented by two similar "Index Catalogues" (IC) that list over 5000 more objects, bringing the total to 13,226, though some are missing. While hardly "new," the NGC and its extensions remain in constant use, rather like Bayer's Greek letters. Many also are the specialty catalogues for diffuse nebulae, galaxies, etc.

Diffuse nebulae are also called emission nebulae as their spectra display bright emission lines (described briefly in Chapter 1) rather than the absorption lines seen in stellar spectra. The story is a bit convoluted. Planetary nebulae (Chapter 6) appear as small disks and rings around dying stars. The Ring Nebula in Lyra (Figure 4.6) is a fine example. In 1864, the English astronomer Sir William Huggins looked at a bright planetary nebula (NGC 6543, yclept the "Cat's Eye") through his spectroscope and found three emission lines, which immediately showed the thing to be made of gas (Chapter 1). A blue emission was quickly identified with hydrogen, while the other two unidentified green lines were hopefully thought to arise from an unknown chemical element, "nebulium" (thus making the discoverer even more famous). Though sixty years later the pair was discovered to be radiated by doubly ionized oxygen (oxygen with two electrons missing), the bright green emissions are still called the "nebulium lines," reminiscent of the "coronium" lines of highly–ionized iron and nickel seen in the solar corona. Finding the same emissions in the spectrum of the Orion Nebula, Huggins proved that it, and by analogy all the other diffuse nebulae, were gaseous as well and not made of faint stars. Modern spectra reveal hundreds of emissions from hydrogen, helium, oxygen, nitrogen, carbon, iron, and so on. Among them, the red hydrogen line and a pair of similarly red nitrogen emissions (from the singly ionized state) usually dominate, the trio giving most diffuse nebulae their reddish colors, though for the core of the Orion Nebula and most of the prominent planetary nebulae, Huggins' green oxygen emissions prevail.

Diffuse nebulae mark the birthplaces of rare stars with masses greater than 10 or so times that of the Sun. With birth temperatures between 25,000 and 50,000 Kelvin, these stars radiate a harsh ultraviolet light so energetic it ionizes the surrounding gases that are left over from stellar formation. A nebula is thus mostly a sea of protons and electrons into which are mixed heavier ions. Densities are typically a vacuous thousand or so protons (and electrons) per cubic centimeter, with some nebulae having far less. As the positively charged ions recapture the negative electrons, the energy absorbed through the ionization process is radiated back in the form of emission lines of hydrogen, helium behaving similarly. While ion-electron recombination also produces fainter lines of carbon, oxygen, neon, and other common elements, the strong green nebulium lines, the pair of nitrogens in the red, and hosts of others are created by ions that absorb energy through collisions with speeding electrons and then radiate it back. They are oddly and collectively called "forbidden lines" (see below). Not truly forbidden as such, atoms and ions are just reluctant to radiate them. The solar coronium lines have a similar origin.

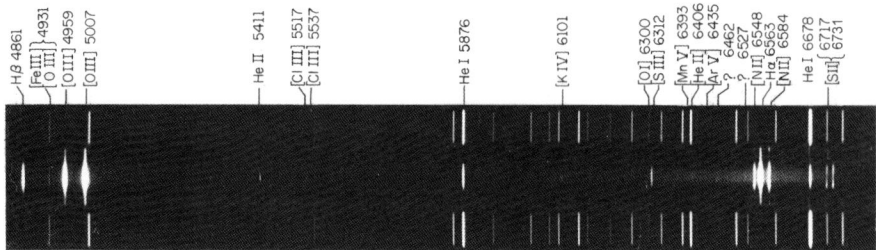

Figure 5.2.   From green at left to red at right, the spectrum of an ionized gaseous nebula (a planetary nebula called NGC 2440: see Chapter 7) runs between a pair of calibration spectra. Two hydrogen lines caused by recombinations between free electrons and protons anchor the ends. In the middle we find similar emissions from neutral and ionized helium (He I and He II). The lines identified within square brackets are forbidden lines caused by collisions between electrons and various species of ions. Particularly prominent are the ionized nitrogen lines that flank the red hydrogen line at far right and the two doubly ionized oxygen lines next to the green hydrogen line at far left. The latter three were those discovered in a similar gas cloud by Sir William Huggins, which proved that such objects are indeed gaseous. Note the wealth of other species that range from neutral oxygen ([O I]) to four-times ionized manganese and iron ([Mn V] and [Fe V]). Lick Observatory spectrogram, J. B. Kaler and L. H. Aller in the Astrophysical Journal.

The secret lives of nebulae

Nebular radiation is worth a more detailed look. The simplest atom, hydrogen, is made of a negative electron orbiting a positive proton, the two tied by the electromagnetic force. Unlike a planet, which could orbit the Sun at any distance, the negative electron can orbit the positive proton only with specific orbital radii at which the electron's orbital energy takes on specific values. The reason for this apparently bizarre behavior is that electrons, protons, and all small particles also behave as waves, and in the classical view you have to fit the waves into the orbits. (In the quantum view, the electron is everywhere in its "orbit" at the same time.) The smallest possible orbital radius for hydrogen, the "ground state" in which the orbital energy is at a minimum, is about half an Ångstrom. A second possible orbit is twice as far out, a third four times as far, then nine times (the ordinal numbers squared), etc. As the orbital radii increase, so do the energies of the bound electrons, but at a progressively smaller rate such that they pile up against a limit. If an electron gains enough energy to top the limit, it escapes, resulting in a hydrogen ion, which is no more than a bare proton.

All physical systems seek their lowest energy states: you are most likely sitting while reading. If not, just like the electron, you will be eventually. In a nebula, neutral hydrogen atoms quickly seek the ground state. But they won't be in it for long. In the nebula, an electron will very quickly absorb an ultraviolet stellar photon that has energy greater than the atom's ionization limit (specifically one with a wavelength shorter than 912 Ångstroms) and be ripped away. After spending what seems like a relative eternity freely zipping along at high speed while exchanging energy with other particles, it's eventually captured by another proton. The greatest chance is that it will go right to the ground state, releasing its energy as another UV photon, albeit at a different wavelength than the one that originally freed it. BUT: it can also land in one of the huge number of possible larger orbits. Say it plops onto orbit 3 (counting outward). Residing only briefly, less than a millionth of a second (the outer orbits are highly unstable), it has a choice of dropping directly to the ground state or briefly visiting orbit 2. If it does the latter, it radiates the difference in orbital energies as a photon with a wavelength of 6563 Ångstroms, in the red part of the spectrum, which helps give the nebula its reddish glow. Capture in orbit 4 followed by a jump to orbit 2 on the way down releases a photon at 4861 Ångstroms in the blue: Huggins' blue line! The final leap from orbit 2 to orbit 1 produces a powerful ultraviolet photon at 1216 Ångstroms. And so on, the choices providing a rich hydrogen spectrum. Those lines featuring level 2 on the bottom are collectively called the Balmer series, those with level 3 the infrared Paschen series, those with level 1 the ultraviolet Lyman series, etc. The lines are named from long wavelength to short with Greek letters, the 3-to-2 transition Balmer Alpha, the 4-to-2 Balmer Beta, the 2-to-1 Lyman Alpha, etc. Since the Balmer lines were the first observed, the red 3–2 is called H-alpha, Huggins's 4–2 H-beta, etc. Heavier atoms and ions with greater numbers of electrons and protons have different ionization limits and far more complex orbital structures through which captured electrons cascade as in a pinball machine. Of these helium dominates. The heavier species must then also have far more complex emission spectra, which along with hydrogen emissions are widely observed in diffuse nebulae from the ultraviolet through the optical and into the infrared, even into the radio domain. (The hydrogen 110–109 orbital transition, "109-alpha," is for example prominent in the radio spectrum and is used, among others, to examine nebulae highly obscured by interstellar dust.)

Excite a transparent box full of low-density hydrogen gas by heating it through electrical discharge or some other means. Most atoms will again be in the ground state, but through violent atomic collisions or by absorbing radiation at a given time a few will have made it to upper orbits. As the electrons drop down, the box radiates emission lines like a diffuse nebula. Shine a light through it, and the electrons in various orbits can absorb its energy, but only the photons with the energies required for the electrons to make it between pairs of orbits. If we then look at the light through the box, we will see the same lines as before, but now in absorption, and we are back to the solar barcode in Chapter 1.

As the free electrons in a diffuse nebula (or a planetary nebula for that matter) dash about, colliding with one another and exchanging energy, they are brought into a velocity equilibrium, allowing a temperature to be defined. The greater the temperature, the faster the average speed of the particles, and the greater the maximum velocity. Like the solar corona, the nebulae are not blackbodies, so the temperature has little effect on luminosity. Nebular temperatures are around 10,000 Kelvin, though if you were actually inside a nebula you would freeze (and with the light spread all around you would see little as well). As the free electrons energetically speed along, they occasionally collide with heavier ions that have special low-lying orbits that, unlike hydrogen's, are remarkably stable. An electron can last in one for minutes, even hours. But the electrons in these quasi-stable orbits have nothing but time, so unless de-excited by another collision, they eventually drop down all by themselves. In simplified atomic theory, the transitions are not allowed, and hence are called "forbidden." When one employs more complex theory, we see that they are indeed allowed, but are just slow to take place. In the lab, the forbidden lines are too weak to see. But there is so much mass in a nebula that they build up great strength, allowing Sir William Huggins and anyone else with a spectroscope or spectrograph to admire them in all their colorful beauty, starting with Huggins's green nebulium lines, which arise from doubly-ionized oxygen.

With the mechanisms in hand, application of atomic theory to the strengths of the emission lines reveal temperatures and densities, and not surprisingly show the element ratios in diffuse nebulae to be similar to those of the embedded stars (though perhaps modified by condensation onto dust grains), which are close to those of the Sun: 90 percent hydrogen, 10 percent helium, and a smattering, under 0.2 percent, everything else. Of the remainder, oxygen dominates, followed by neon, carbon, and nitrogen, with abundances rapidly dropping off toward the heavier elements. Faintly glowing red from the hydrogen and forbidden nitrogen emissions, diffuse nebulae are set against a vast scrim of much thinner ionized clouds that pervade the Milky Way. One is hardly ever free of them.

Mirror mirror

Long-exposure imagery shows the Pleiades (the Seven Sisters in Taurus; Figure 4.9) to be enmeshed within a filamentary nebula that concentrates around the cluster's brightest star, Merope. Unlike the reddish (and greenish) Orion Nebula the cloud around the Pleiades shines distinctly blue. Moreover, there are no emission lines in the nebula's spectrum. Instead, we see an absorption spectrum that matches those of the brightest embedded stars. The cloud around the Pleiades is an example of a "reflection nebula" caused by grains of interstellar dust that scatter (in a loose sense reflect) starlight. The scattering efficiency rises strongly with decreasing wavelength. Moreover the stars of the Pleiades are also hot and bluish themselves, resulting in the dramatic blue haze in which they are embedded. Diffuse nebulae also have reflection components, showing that interstellar clouds consist of both gas and dust. The difference between diffuse and reflection nebulae comes down to the temperature (and ultimately the mass and age) of the buried star or stars. If one or more are sufficiently hot to produce enough energetic ultraviolet radiation that can ionize the surrounding gaseous cloud, we see a diffuse nebula. If the star or stars are not hot enough, the gas remains neutral and effectively unobservable. What we observe then is the dust component as the little grains scatter the starlight. Some nebulae, like the striking Trifid Nebula in Sagittarius (M 20), simultaneously show evidence for both kinds of nebula, blue reflection nebulae beautifully set against red diffuse clouds. In a few cases, the reflection nebulae are created by reddish stars and appear more yellowish than blue. The huge nebula surrounding the great supergiant Antares in Scorpius is an example.

Bumping through the darkness

The Pleiades are (is?) passing through a dusty interstellar cloud that does not relate to the cluster's birth. When it leaves the cloud behind, with no illumination the blob of interstellar matter will become a "dark nebula" whose dust will (as it does now) slightly dim the brightness of the background stars. Some dark clouds, like the Coalsack next to the Southern Cross (Chapter 4), are so thick that they block the background entirely, appearing like holes in space. They are not to be confused with "black holes" with which they share only the word "black." Nor are the dark opaque clouds to be confused with "dark matter," as they are made of ordinary protons and electrons like the rest of us. As much as anything, "dark" in "dark matter" means "unknown." We know what the dark interstellar clouds are made of. Moreover, they shine brightly at longer infrared and radio wavelengths where they radiate away their absorbed energy. Also called "globules," conglomerations of the dark clouds create the Great Rift that splits the Milky Way in two and that give the starry stream much of its character and beauty. The narrow rift shows that the dust makes the thinnest layer of the galaxy's disk.

Figure 5.3. Thackaray's Globules are thick, dark, dusty molecular clouds starkly set against a more distant ionized background, the red color coming from the hydrogen and ionized nitrogen emissions seen in Figure 5.2. Globules and molecular clouds that range from small ones like these a bit over a light year across to giant clouds a hundred times larger are star-formation factories. The dust keeps out heating starlight, rendering the clouds cold enough to allow condensation of the dominating molecular gas, which is mostly hydrogen. NASA and the Hubble Heritage Team (STScI/Aura).

While interstellar dust is both necessary to star formation and interesting in itself, its dimming effect is remarkably annoying. Within or even close to the Milky Way, nearly all stars at any substantial distance appear fainter than they actually are. The effect is vaguely similar to the scene on a foggy night while driving in the country. If all goes well, you can get a star's distance by estimating its absolute brightness from its spectrum (Chapter 6) then comparing it with its observed brightness. Dimming by dust that lies in the line of sight, however, will make the star look too far away. Fortunately, interstellar dimming gets much stronger toward bluer wavelengths, the complement to the increased scattering efficiency that makes reflection nebulae look blue. Starlight is thus also reddened in some proportion to its absorption. From the kind of star (Chapter 6), we know the real color (which is technically defined by comparing magnitudes measured at different wavelengths). Comparison to the observed color then gets the amount of absorption and the true apparent brightness (that seen with no dust) and, depending on the technique used, the distance. Providing that the ratio of reddening to absorption is correct, which it often isn't, leading to yet more screams of frustration.

The absorption of starlight in the plane of the Milky Way is so severe that in the visual spectrum, we can't see through it, and no other galaxies are visible at all, creating a "zone of avoidance" that was most confusing until we understood that the "spiral nebulae" were really external star systems like our own. There is so much dust in the line of sight to the center of our galaxy, 25,000 light years away, that it is completely blocked in the visual spectral realm by 30 magnitudes of extinction, that is, only one visual photon in a trillion gets through.

How can we then observe the galaxy's center, know anything about it? The secret word is "radio," first used for astronomy in 1933 and fiercely developed in the mid-twentieth century (Chapter 3). Long-wave photons in the radio spectrum, even those in the long-wave domain of the infrared, are not much bothered by the dust. You have no more problem listening to broadcast radio on a rainy or foggy day than you do when it is clear and sunny. Radio waves are too long to "notice" the pesky dust grains and simply ignore them. More important here, radio radiation allows us to look into the hearts of the dark dust clouds to see what is going on. Dark they may be in the visual realm, but in the radio they radiate brilliantly and are seen as bright emission nebulae, the emissions coming from hydrogen, helium, and most importantly even hosts of molecules. The infrared, while subject to some absorption, is similar.

The interstellar medium is heated in part by the ultraviolet radiation of hot stars. While the temperatures of diffuse nebulae hover around ten thousand Kelvin, in a thick dust cloud with no internal or external source of energy, one like the Coalsack perhaps, the dust absorbs the energetic radiation from the galaxy's hot stars, shielding the interior cloud, which then turns cold. Temperatures fall to not much above absolute zero (−273 degrees Celsius, −459 degrees F).

The gas in the clouds is then made mostly of molecules (helium always in the atomic form). Molecules are fragile and rapidly broken by collision, the reason there aren't any in hot stars, nor even many of them in the Sun. And in a hot diffuse nebula, the atoms and ions collide much too harshly for them to join to create molecules. But in a dark nebula, the atoms can catch on to one another. The result is a cold chemistry that is aided by interactions between atoms on the surfaces of dust grains and by the passage of cosmic rays, which are high-speed atomic particles from exploding stars (Chapter 7) that can create ions and thus also help in the chemical reactions. Unlike laboratory chemistry, which is generally heat-driven and fast, most dark cloud reactions are terribly slow. But again we have nothing if not time. As a result, the clouds build up an amazing chemical morass that has profound implications even for life on Earth.

Discovery of the first interstellar molecules, CH (carbon and hydrogen) and CN (cyanogen: carbon and nitrogen), which appear in absorption in optical stellar spectra, goes back nearly a century. "Hold on here" say the chemistry majors, and even perhaps those who flunked the subject. "These aren't molecules, they're radicals, incomplete molecules, to which another atom would very rapidly attach itself." But densities are so low even in thick interstellar clouds that radicals have great stability (there's not much around to attack them) and, so far as astronomers are concerned, they're true molecules.

Interstellar molecular radio astronomy took flight in 1963 with the discovery of the OH (hydroxyl) radical. OH radiation commonly behaves as a powerful "maser," the microwave (short-wave radio) version of the well-known laser. "Maser" stands for "microwave amplification by the stimulated emission of radiation." Masers and lasers ("light amplification...") are created when there are more atoms and molecules with electrons in upper energy levels than in lower ones, the reverse of the usual case. When triggered, downward dropping electrons stimulate others to cascade downward as well, producing an ultra-powerful beam of radiation. The stunning discovery of OH was followed by those of water, carbon monoxide, ammonia, alcohols (both methyl and ethyl), formaldehyde, ethers, cyanides, the list going on and on to hundreds, some of the others (water, silicon monoxide) masering as well.

Among the odder are molecules made of long carbon chains onto which are attached nitrogen and hydrogen atoms. They are so fragile that they don't exist on Earth. But in the cold low-density environments of thick interstellar clouds, they can survive, the clouds thus becoming natural laboratories. Two of the most intriguing species are acetic acid and urea, which begin to relate to

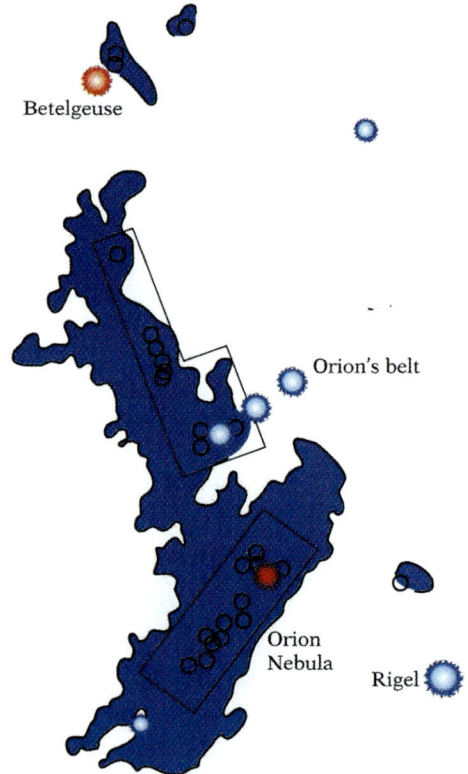

Figure 5.4. The vast Orion Molecular Cloud (combined northern and southern clouds) lies in back of the eastern part of the constellation. Visible with radio telescopes, it's one of the more active star-formation regions known. The Orion Nebula, in red, is an ionized cavity at the cloud's front edge caused by recently-formed hot massive stars. From "Cosmic Clouds," J. B. Kaler, Scientific American Library, Freemen, New York, 1997, adapted from R. J. Maddalena *et al.* in the Astrophysical Journal.

biological molecules like glycine and to other amino acids that are ardently sought. Could the clouds contain the seeds of life? The molecules become so large and complicated (benzene rings attached to one another, "bucky balls" made of 60 or more carbon atoms) that they begin to behave like tiny interstellar dust grains, thus completing a circle. Many spectral emissions (and even interstellar absorptions, which flock the ultraviolet) remain unidentified, in part because laboratories on Earth cannot reproduce them and we must use atomic theory.

In spite of all the molecular complexity, most of the mass of the dark clouds is in the form of simple molecular hydrogen, H_2, made of two hydrogen atoms. A molecule made of a pair of the same atoms does not have strong radio emission and is difficult to observe, so carbon monoxide is commonly used as a tracer to map out what are now known as "molecular clouds," big ones as "giant molecular clouds" ("GMCs"), which can contain thousands, even millions, of solar masses and are the true birthplaces of nearly all the different kinds of stars. They are accessible to the naked eye on a dark, moonless night with no artificial lighting. You need only to gaze at the Milky Way and its extraordinary structure.

Birth

Look at the remarkable symmetries of the planetary system (Chapter 2). All the planets orbit the Sun in the same direction close to a plane (the ecliptic) not far off the Sun's rotational equator. Aside from Venus and Uranus (and Pluto if you still want it as a planet), they all rotate in the same direction as well, with their rotation axes more or less perpendicular to their orbits. Even backwards Venus has its axis nearly straight up and down. Moreover, the larger moons (particularly our own Moon, the big four of Jupiter, and Titan of Saturn) partake of the same orbital directions and orientations. The symmetry is so obvious that in 1755 the great philosopher Immanuel Kant (1724–1804) suggested that the planets were formed from a gaseous flat "cloud" (a nebula) that once surrounded the Sun. In 1796, the mathematician and astronomer Pierre-Simon Laplace (1749–1827) took up the concept as well, that the planets were formed from a spinning nebulous disk. And there the "nebular hypothesis" languished for two centuries while other theories came and went. However since the planets are satellites of the Sun, we can get further insight by looking not just at the formation of the planets, but at the formation of stars in general. Then we might also speculate as to the existence of other planetary systems, to how we might observe them, and about what roles they might play in understanding ourselves.

Stars have finite energy sources, so they must all eventually die. Yet a look overhead shows countless numbers of them, so they somehow have to be replaced. We can't see it happening, but there are all those dark, opaque clouds... they must be the birthing rooms! All we need to do is to look inside them, which had to wait for the development of radio and infrared astronomy. Even in the visual spectrum, however, there were some clues. Ninth magnitude T Tauri, found in 1852 near the Hyades cluster, varies erratically in brightness. Its spectrum shows that it is both accreting matter from its surroundings while at the same time ejecting gas in a strong wind. Moreover, it seems to be in a high state of solar-type activity, implying fast rotation. It's near the fringe of a dark cloud and illuminates a reflection nebula, Hind's Variable Nebula, named after its discoverer. Moreover it's at least double. Thousands of similar "T Tauri stars" are now known, all associated with dark clouds, some of the stars wildly variable.

Meanwhile, in the middle of the twentieth century George Herbig (1920–2013), then of Lick Observatory, and the Mexican astronomer Guillermo Haro (1913–1988) began to find strange small, fuzzy nebulae that radiated emission lines yet had no seeming sources of energy, no illuminating central or neighboring stars. For a time we thought that Herbig–Haro ("HH") objects were actually stars in the act of formation. But further observation revealed that they come in well-separated pairs each centered on a T Tauri star. A closer look yet revealed jets that not only emerged from the T Tauri stars but, spanning light years, pointed directly at the HH objects. Sometimes there would be but one jet and HH object, whereupon it was readily assumed that the other was hidden by the dark cloud that bordered the star. What could focus the jet? A disk! The strong wind from the T Tauri star could only emerge from the disk's poles where there is little resistance, perhaps aided by magnetic fields formed from the combination of spinning disk and rotating star. The new star is simultaneously accreting matter from the disk and ejecting some along its poles while at the same time planets are probably forming in the circulating disk. We seem to be seeing the nebular hypothesis of Laplace in action. Planets have to be a natural part of star formation, and we really can't separate them out.

Figure 5.5. Hubble's "Pillars of Creation" made international news. Some five light years long, these "elephant trunks" within the Eagle Nebula (Messier 16) in the constellation Serpens are filled with new stars. As ultraviolet radiation from older hot stars boils away the pillars' tips, the newcomers see their first "light of day." The Sun may have been born in a similar structure. NASA/ESA and the Hubble Heritage Team (STScI/AURA).

Everywhere within and around the interstellar clouds we see evidence for star formation. Heat a gas and it expands (the fundamental idea behind a steam or gasoline engine); chill it and it contracts. It's only within the dark cold clouds then that stars can condense. A molecular cloud is far from smooth. It's shredded by collisions with other clouds, compressed and torn by hot-star radiation and winds, and rocked by shock waves from nearby supernovae (exploding stars). A shock wave is formed when a disturbing particle moves faster than the natural speed of sound. Examples are the bow wave off a speeding boat and a sonic boom from a supersonic aircraft. The "boom" is not the "breaking" of the sound barrier but is a continuous over-pressure in the air that follows along after the craft. And yes, as long as there is a transmitting medium there is "sound" in space, even if it could not be heard by a human, its speed and intensity dependent on the medium's temperature and density. And there is always a transmitting medium, no matter how thin. Even in the space between galaxies (which contains a hot low-density gas that radiates X-rays).

The result is that the molecular clouds are lumpy, filled with dense blobs. If one gets thick enough as a result of such violent stirring, it can begin to contract under its own gravity to eventually form a star, or if somehow subdivided a double or multiple star. If the cloud is large enough, many bloblets will contract to form a cluster that might someday look like the Pleiades or Hyades. A stellar birthing cloud must be rotating because of collisions and gravitational interactions. As a result we have to consider the conservation of angular momentum (it's back), which first encountered in Kepler's second law of planetary motion (Chapter 2). As a singular blob of gas and dust collapses, like an interstellar skater or dancer it rotates faster and faster, enough to tear itself apart, thus showing that stars cannot form, reminiscent of the old saw that "science shows bumblebees cannot fly." But fly they do.

And form they do. So something must slow the rotation. The major process seems to involve magnetism. Our galaxy is rotating around the galactic center. More accurately, its individual stars (and the Sun) as well as the interstellar clouds, are in collective orbits around the center that make the galaxy appear to be rotating en mass. With a rotational speed of 220 kilometers per second and a distance from the galactic center of 25,000 light years, it takes the Sun some 220 million years to go all the way around. Given that the Earth and Sun are 4.6 billion years old, we've made the trip 20 times. We are on a speeding spaceship, and the scenery — the set of constellations made from local stars — is constantly changing, though over the course of written history it's not noticeable to the naked eye.

The interstellar medium is also filled with ions that reside in diffuse nebulae and in vast near-vacancies blown out by exploding stars. With temperatures of hundreds of thousands of degrees, the voids are filled with highly ionized heavy atoms, such as oxygen, that radiate X-rays. We in fact are on the edge of one created by the deaths of massive stars in Scorpius and Centaurus. Move an electrified particle and you produce a magnetic field. The Earth's field is created by circulating molten iron in its outer core, while the Sun's is made by rotation and convection of its hot gases. Move the ionized matter of the whole galaxy, and lo, you have a galactic magnetic field.

What scientific Sherlock could detect such a field? Atomic energy levels, like those described above for hydrogen, are split by magnetism. As a result, the related absorption or emission lines are also split, or at least broadened, in proportion to the field's intensity, allowing a measurement of magnetic strength. The "Zeeman effect" (after the Dutch physicist Pieter Zeeman, 1865–1943) is readily detected in the Sun, especially in sunspots, where the field strength can reach thousands of times that of the Earth's. In certain magnetic stars, the fields can hit millions of times the terrestrial value. (Have we mentioned that the Earth's field has been decreasing? We may be in for one of the periodic, million-year reversals in which the field can disappear, leaving us vulnerable to solar flares and coronal mass ejections. But we digress…)

Using sensitive radio receivers, the interstellar galactic field is measured to be about a millionth that of Earth's. While that seems small, our galaxy is nothing if not big, so big that galactic magnetism takes up a quarter or so of the galaxy's energy content. The field threads its way through a contracting dark pre-stellar blob. Filled mostly with a cold molecular gas, the cloud is largely neutral and thus ignores the magnetic field. However, the galaxy is also filled with high speed particles, cosmic rays, mostly protons, accelerated to nearly the speed of light by exploding stars. The cosmic rays are constantly zipping through the blob, leaving a trail of ions (the same ones that promote cloud chemistry) behind them. The local field grabs on to the ions, the ions bump into the neutral molecules and atoms, and the cloud gradually slows its rotation enough to contract under its own gravity. Or so we think.

You are sitting in cosmic ray showers as you read. Accelerated by exploding stars, and confined by the galaxy's magnetic field, protons, as well as heavier nuclei and electrons, crash steadily into the Earth's atmosphere. With their immense energies, they break ambient atmospheric atoms into smaller particles that rain downward. They are a part of your natural radiation background. To an astronaut in deep space they are dangerous. The most energetic cosmic ray ever detected, the "oh my god" particle, had the energy of a professionally-pitched baseball. Cosmic ray collisions create the carbon-14 that is used in the dating of organic materials and may influence, even cause, rain showers and thunderstorms. We are protected from their direct impact by our atmosphere and magnetic field. However, the Earth's field is weakening and perhaps may disappear for a while.

The pre-stellar blob's contraction still cannot be brought to a halt. As its core contracts, it heats, and the forming star eventually brightens enough to be seen. It has also been rotating yet faster, so the stuff that has not readily fallen into the star gradually flattens into a disk. Developing its own magnetic field that wraps around the collapsing core, the blob begins to shoot jets outward perpendicular to the disk (the direction in which the disk is thinnest), which removes yet more angular momentum. The outbound jets sweep up ambient gases, shocking them into Herbig–Haro objects. At the same time, the core's dense center heats up enough inside, into the millions of degrees, to the point where it can sustain thermonuclear (hydrogen to helium) fusion. With a new energy source balancing gravity, the contraction slows and halts, allowing a true new star to be born.

The vast majority of stars are of low mass, far below that contained by the Sun, perhaps all the way down to what we might call big planets. The greater the mass, the fewer blobs there are until somewhere above 100 solar masses, no stars are made at all. If greater than ten or so solar masses, the new star or stars are so hot at their surfaces that they radiate enough energetic ultraviolet light to ionize their surroundings, and a diffuse nebula is born. If not that hot, maybe the new star will create a reflection nebula. Whatever the case, nebula or not, high or low mass or in the middle, the dark cloud gradually dissipates, victim to star formation and stellar winds, while the stars that it birthed gradually move away to orbit within the galaxy long after they have run their course and die as celestial cinders, glowing forever dimmer.

Planets

We are left with a new star nested within a surrounding residual disk. Astronomers now have the capability to see the disks posed so long ago by Kant and Laplace. As the dark clouds of the galaxy birth its stars, the dusty disks around them should be the birthplaces of planets. The tiny grains within the disk bump into one another and stick together, small grains turning into bigger ones. Further encounters produce rocks, then larger "planetesimals," millions, billions of them. At this point they begin to have significant gravitational fields, allowing the bigger ones to attract the smaller (though many of the important details of the process are still obscure).

Within a mere one to ten million years, the disk has created planets, or at least planetary cores. Back home, in the primitive inner Solar System, the heat from the Sun was so great that new planets, what will be Mercury through Mars, cannot accrete volatile elements and molecules such as water. Beyond about five AU from the Sun, however, we meet up with the "snow line," where the temperature is so low that the growing cores can accrete not only water, but the most abundant of substances, hydrogen and helium. Jupiter and Saturn thus grow fat, Saturn the smaller of the two because there is less matter in the disk to work with. Farther out, where the disk thins even more, it creates the lower-mass, higher-density planets Uranus and Neptune.

Competing planets merged. The Moon was most likely created by a collision between Earth and a Mars-sized body that had the misfortune to cross our orbit at the wrong time. Some of its mass joined with Earth, while the rest of the debris condensed to form the Moon. The concept explains the angular momentum of the Earth–Moon system and some of the bulk chemistry, but unfortunately not its isotopic ratios. The heat of formation, especially collisions among larger bodies, partially melted the inner planets and caused heavy elements like iron and nickel to fall to the planetary centers, giving the terrestrial planets, even the Moon, metallic cores. Between Mars and Jupiter a planet perhaps tried to form, but Jupiter kept the growing bodies stirred up, resulting in the colliding, broken asteroids. The best they could do was 1000-kilometer-wide Ceres. Of lesser mass we find the rest of the named and/or numbered asteroids (nearly a million of them) and billions of smaller particles. Out beyond Neptune, where there was a dearth of icy matter, no significant planet could coalesce, and we are left with Pluto, Eris, and the rest of the Kuiper Belt. The outer planets threw huge numbers of icy junk, nascent comet nuclei, out into what became the Oort Cloud. Once the inner planets solidified, they were bombarded by asteroidal and cometary remains that invaded the inner Solar System thanks to the gravitational perturbations of the outer planets, resulting in intense cratering, which we still see on the Moon, Mercury, and Mars. Earth was hit too, which early-on may have brought our water, but erosion and tectonic processes wiped the early craters away. The various large satellites of the outer planets were perhaps formed from rings of debris that settled around them. These suffered from heavy bombardment too.

That is the standard, simple view. Unfortunately, a lot of it is wrong. Modern formation models increasingly show that gravitational interactions among the planets and with the remains of the Sun's accretion disk, including the stuff in the Kuiper Belt, caused the giant planets to migrate within the Solar System, Jupiter perhaps approaching the current orbit of Mars and destroying any early asteroid belt. When it moved back to where it is today, it dragged inner debris outward to make the current asteroid belt, while Uranus and Neptune moved outward, perhaps doubling their distances from the Sun, Neptune shoveling the new Kuiper Belt, including Pluto and the like, in front of it. Evidence for such planetary migrations is written into the varied planetary systems being discovered orbiting around extraordinary numbers of stars.

Passing stars and molecular clouds still disturb the icy bodies of the Oort Cloud, throwing occasional long-period comets to us, while migrations to the inner planetary system from the Kuiper Belt produce the short-period comets. Comets and asteroids can be big enough for their collisions to wipe out whole species of terrestrial life (dinosaurs for example). Small pieces, shards that range from large rocks to sand grains, produce occasional meteors as they strike our atmosphere. Rare bigger ones give us brilliant fireballs that, overheated, sometimes explode. Comets disintegrating under sunlight generate dozens of annual meteor showers as well as most of the general meteors of the nighttime sky. Looking back at our system from outside, an alien astronomer would see the disk of planets embedded in a thin dusty disk left over from formation and added to by comet dust and the remains of asteroid collisions. And might even wonder about the third planet from the Sun, which is at the right temperature to sustain liquid water. We will meet these and others in our story's last chapter.

6

An Infinite Variety

Well, not really "infinite." That would imply a cosmologically infinite Universe. Since there are only so many parameters you can use to define a star, if the Universe is infinite (for which there is no actual evidence) there must a duplicate of the Sun (or any other star), indeed an infinite number of them! (Credit "Cosmology," R. R. Harrison). Here "infinite" is an apt metaphor for the immense array of stars that nature creates from the dark dusty clouds of interstellar space, of which the tour in Chapter 4 and the introduction to star formation in Chapter 5 give but a hint. (On the other hand we live in a quantum Universe where particles act like waves and cannot be pinned down in both velocity and position at the same time: the Heisenberg Uncertainty Principle. That could mean that even in an infinite Universe true replication is impossible.)

As for bugs and elephants, the first step in understanding stars is classification, a dreaded word that makes people leave the room. To the contrary. Not only are the similarities and differences among the stars of interest on their own merits, it's through classification that we can open up stellar natures and string together the stars' life cycles, from their births in the last chapter to their oft-remarkably beautiful deaths in the next. We had little clue about stellar kinds, just colors, until we could look at the stars' spectra, which consist of colored backgrounds upon which are superimposed dark lines, properly put, absorption lines. Running perpendicular to the flow of colors, the lines range from miniscule depressions in the spectrum to wide valleys in which nearly all the light is removed over a broad band of wavelengths. First approached in Chapter 1 with regard to the Sun, the absorption lines, which look like store barcodes, are the reversals of Chapter 5's emission lines. Here the electrons tied to atoms, ions, and molecules, are raised from lower to higher orbits by absorption of light with energies (hence wavelengths) that correspond to energy differences among allowed states. While there are but four lines from hydrogen in the visual (red through violet) spectrum, there are millions over the spectrum ascribed to metals, which have frighteningly complex electronic structures. The interior of the star provides the background continuous (no breaks) blackbody light (Chapter 1), while the cooler outer skin, or thin "atmosphere" as it is called, superimposes the absorption lines. If you look at the outer stellar atmosphere alone, as we can do during a solar eclipse, you see the emission lines of Chapter 5. Our thanks to the German physicist Gustav Kirchhoff (1824–1887), who first elucidated the laws by which different kinds of spectra are created: continuous (from blackbodies with no lines at all), the emission lines of Chapter 5, and the absorptions first noted in Chapter 1 and examined in much more detail here.

By their spectra shall ye know them

The first foray into the mysterious wilds of stellar spectra was made by the English physicist William Wollaston (1766–1828), when in 1802 he discovered absorption lines in the solar spectrum. He erroneously thought that their purpose was to divide the colors. A dozen years later, the German optician Joseph von Fraunhofer (1787–1826) mapped hundreds of absorption lines. Not knowing their origins, he assigned Roman letters to the strongest, starting from red and working to violet, using first upper case and then lower case. Laboratory experiments by Kirchhoff and the German chemist Robert Bunsen (1811–1899, of Bunsen burner fame) began to reveal their chemical origins, putting the lie to French philosopher Auguste Comte's (1798–1857) famed comment that we could never know anything about the chemical compositions of the stars. To the confusion of generations of students, the Fraunhofer designations, which are still in common use, have nothing to do with the lines' chemical identifications. In the red, Fraunhfer A is produced by molecular oxygen in own atmosphere (through which sunlight obviously must pass), the b lines are from neutral magnesium, the two closely-spaced yellow D lines with wavelengths around 5900 Angstroms are made by neutral sodium, while ultraviolet H and K, which are by far the strongest of all solar absorptions, arise neither from hydrogen nor potassium but from ionized calcium. You rapidly get used to it. (We can distinguish solar lines from those superimposed by the Earth's atmosphere as the latter get stronger as the Sun sets and sunlight passes through a greater thickness of air.)

The first observations of stellar spectra in the middle of the nineteenth century (which were through the telescope eyepiece, not via photography) were puzzling. While some spectra looked rather like that of the Sun (Capella for example), others were completely different. The H and K absorptions in the spectra of Vega and Sirius were seen to be very weak, while the hydrogen lines, prominent in the solar spectrum but far from strongest, are overwhelmingly broad and powerful. Then we look at reddish stars like Betelgeuse, Antares, and especially the red variable star Mira, and find hydrogen to be absent, replaced by "fluted bands" that later turned out to be from the titanium oxide molecule, TiO. It looked as if stars were made of different elements in various proportions, but nobody really knew.

All the astronomers of the time could do with stellar spectra was to come up with some kind of scheme to classify them. Placed in order, the spectra might then make some sense. A number of false starts culminated in a turning point developed in the 1840s by the Jesuit astronomer Angelo Secchi (1818–1878), who separated the spectra into four (later five) groups. With bows to Secchi's scheme, and with the advantage of permanent recording by photography, E. C. Pickering (1846–1919) of the Harvard College Observatory got the stellar idea of classifying stars mostly on the basis of their hydrogen lines, "A" for those with the strongest hydrogen lines, "B" for stars with weaker hydrogen lines, and so on down to "Q," the criteria subjective but defined by particular stars. Adding to potential confusion, the letters have nothing to do with either the chemical

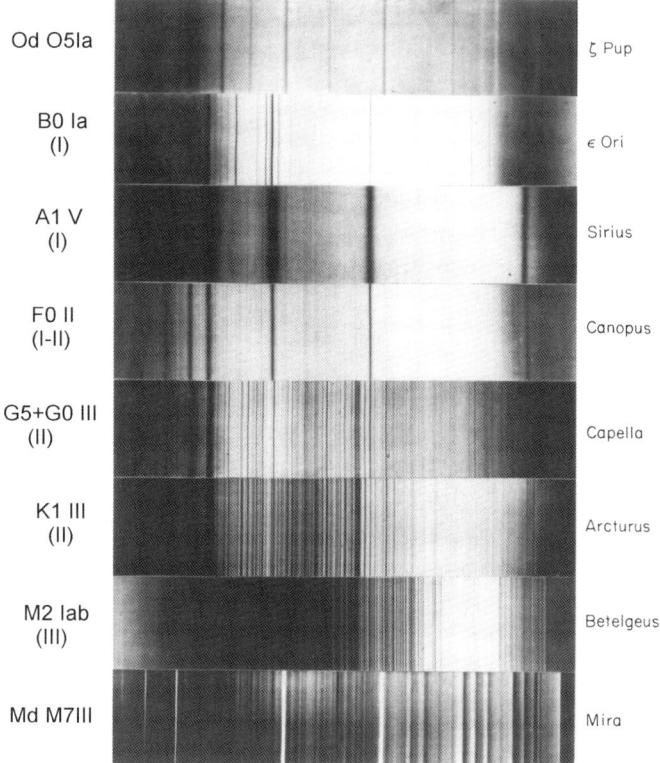

Figure 6.1. The spectral sequence from class O to M is presented as originally conceived. In spite of the enormous variation from hot (top) to cool stars, they are all made of the same thing: 90 percent hydrogen, ten percent helium, and a smattering of everything else. All the variation is caused by temperature. Hydrogen absorptions, which along with others are created at the fuzzy stellar surfaces, are maximized in class A. Toward cooler stars they weaken as fewer atoms become capable of absorbing outbound stellar radiation until they disappear in class M. As hydrogen weakens, absorptions by metal atoms and ions increase, giving the spectra greater complexity. Class M stars are so cool that they are dominated by even more complex molecular absorptions. Hotter than class A, hydrogen becomes ionized, leaving progressively fewer neutrals to absorb the outbound radiation. Helium absorptions are seen only in hot classes B and O. From the Annals of the Harvard College Observatory, Vol. 28, Pt. II, 1901.

identifications or with the Fraunhofer designations. "K stars" are not rich in potassium, but have relatively weak hydrogen lines, one giant step down from G stars like the Sun, where they are notably stronger.

From among the observatory assistants, three stood out: Annie Cannon, Antonia Maury, and Williamina Fleming. The trio largely took over the classification task. Along with Pickering (without being specific as to who did what and how they did it), they dropped some classes as erroneous or redundant, then re-arranged them to account for continuity among other absorptions, in particular those that eventually turned out to be from helium. The result is the fundamental, classical, alphabet of stellar astronomy into which most stars fall: OBAFGKM. Seven bins were really inadequate for the detail that was being observed, so Cannon and her crew decimalized them, the hydrogen lines at F0 stronger than those at F5, other absorptions falling into place. By 1930, Annie Cannon had classified 250,000 stars, by 1945 350,000, an amazing achievement. Classifications, celestial positions, and brightness estimates, are all available in the Henry Draper Catalogue, which was financed from Dr. Draper's estate and is still viable.

And lo! Early on the spectral sequence was seen to correlate with star color, O stars blue, B and A white, FGK progressively more yellow then yellow-orange, M stars reddish. (Washed out and subtle, star colors as viewed with the human eye are subjective, hence contentious, various people seeing them differently. The only way to express stellar color properly is via comparison of magnitudes at different wavelengths, as discussed in Chapters 4 and 5.) Low temperature stars can emit only lower energy photons, and are thus reddish to the eye. As temperature increases, higher energy photons are released and the apparent color of a radiating body changes, going through those of the rainbow, becoming progressively orange, yellow, white (subbing for green), and then blue (see Chapter 1). It was thus brilliantly clear that the spectral sequence is also a temperature sequence that runs from hot and blue class O near 50,000 Kelvin through white A stars like Vega near 10,000 K, past the class G2 Sun at 5780 K, on into class M in the 3000 K range. The first thing that any astronomer wants to know about a star is its class.

In the 1920s, the English-then-American astronomer Cecilia Payne-Gaposchkin (1900-1979) showed that the wild changes that appear in the spectra along the sequence are not caused by differing chemical compositions, but by increasing ionization and efficiencies of absorption as one climbs the temperature ladder. The hydrogen lines start off fairly weak at higher temperatures, peak with great strength at about 9500 Kelvin, then diminish to near-nothing at the lowest M-star temperatures. Hot O stars are known for their ionized helium, B stars for their neutral helium, then from class A on down we see a diminishment not only of hydrogen lines, but of metallic ionization and general atomic excitation. Class M hosts molecules (particularly titanium oxide), which are broken up at higher temperatures by collisions among ever-faster-moving particles. As an example of what is going on physically, the H and K lines of ionized calcium arise from the ground orbital state, so virtually all the ions can act to produce them. The Balmer absorptions of hydrogen, however, all arise from the second orbit outward, which is only sparsely populated among the atoms by atomic collisions. At any given time, only one hydrogen atom in a hundred million is capable of Balmer absorption. So even though there are 10 million hydrogen atoms to one of ionized calcium, the H and K lines come out stronger. The actual chemical ratios can be calculated by atomic rules worked out in the 1920s, and most are similar to those found in the Sun. The basic concept is that the various energy levels are populated by a combination of atomic collision and radiative absorption and emission. Once the various probabilities of energy level populations and transitions are known from the lab or theory, atmospheric parameters (temperature and density) as well as chemical abundances can be found. We are still working on it.

White class A stars are all over the sky, Sirius and Vega (respectively A1 and A0) ascendant. But there are also Deneb (A1) and Altair (A7), which with Vega form the bright Summer Triangle. Then look to the Big Dipper, whose middle five stars range from A0 to A5. Dropping down, at F5 Procyon of Canis Minor (Orion's smaller hunting dog) lies in the middle of the class, while the best-known star of all, Polaris (oddly in Ursa Minor), is F7. If you want a

Table 6.1. The spectral sequence.

Class	Spectrum	Color	Temperature (K)
O	Ionized and neutral helium, weakened hydrogen	Bluish	31,500–49,000
B	Neutral helium, strong hydrogen	Blue-white	10,000–31,500
A	Powerful hydrogen, ionized metals	White	7500–10,000
F	Weaker but still-strong hydrogen, ionized metals	Yellowish-white	6000–7500
G	Still weaker (but still strong) hydrogen, ionized and neutral metals. Strong H and K lines.	yellowish	5300–6000
K	Relatively weak hydrogen, neutral metals, stronger H and K.	Orange	3800–5300
M	Little or no hydrogen, neutral metals, molecules (TiO and VO)	Reddish	2100–3800
L	No hydrogen, metallic hydrides, alkalai metals	Red-infrared	1200–2100
T	Methane	infrared	Under 1200
Y	Ammonia	infrared	300?

classic G star, just look (metaphorically, not really) at the G2 Sun, while Capella in Auriga, a double star (more about these later) with G0 and G8 components, spans the range. Vying for naked-eye popularity are the K stars. You can find them easily by their subtle orangish colors. The front bowl stars of both the Big and Little Dippers, Dubhe and Kochab, at K0 and K4, fall into the class. Indeed, much of Ursa Major (the Greater Bear that holds the Big Dipper) is made of K stars. Toward the cooler end we find K5 Aldebaran, which lies in front (but is not a part) of the Hyades cluster of Taurus. Two reddish zeroth-magnitude class M1.5 stars lie opposed in the sky, Betelgeuse of Orion and Scorpius's Antares. Much deeper into the M stars is the marvelous long-period variable Mira near the heart of Cetus, introduced in Chapter 5's tour. Typically M7, Mira varies in brightness over a 330-day period, as does its surface temperature and class. While it can hit second magnitude, most of the time it's invisible to the naked eye. Hotter than class A, B stars abound, most of them flocking the Milky Way, though there are many exceptions. Perseus, Scorpius, Lupus, and Orion (with bright B8 Rigel and, near the top of the class, B2 Bellatrix) contain truckloads of them. Others of note include the luminary of Leo, Regulus (B7), Alkaid (B3) at the end of the handle of the Big Dipper, all the naked-eye Pleiades, and the B1-B4 double, Spica, in Virgo. Going still hotter into the O stars gives us a bit of a problem, as they are extraordinarily rare. Among the few naked-eye examples are O8 Lambda Orionis, which with others forms Orion's head, and fifth magnitude 10 Lacertae (O9). Hotter is O6 Theta-1 Orionis C, which powers the Orion Nebula. Here is one of the Great Divides. Only class O stars and those at the very high end of class B (B0 etc.) are hot enough to fire up the diffuse nebulae and are massive enough to explode. We'll be back to them and the Great Divide later.

The weakening of the hydrogen lines among stars hotter than class A might seem a bit curious. As temperature rises from the bottom of the sequence, collisions among atoms increase in energy, and the electrons of more and more atoms are kicked up into the second orbit. The Balmer lines should then continuously increase in strength. But above 10,000 or so Kelvin, into the B and O stars, atomic collisions become so violent that the electrons are removed altogether, and the hydrogen becomes ionized. The result of the ionization is a free electron and proton. Since the proton now has no electron, it has no spectrum. And since there now are progressively fewer neutral hydrogen atoms as temperature climbs, the Balmer lines decrease in their absorptive power.

But wait, there's more… Just when we get a handle on the seven classic types, we find that we're only part way through our taxonomic journey. Red carbon stars, traditionally class N, overlap the temperature range of the M stars, lesser-used "R" overlapping much more common G and K. The two are now combined into "C." In normal stars like the Sun there is twice as much oxygen as carbon, while in carbon stars, the ratio is reversed, for some profoundly so. Discovered around 1845 by Father Secchi, whose classification work augured the modern spectral sequence, their low temperatures plus strong absorption bands of cyanogen (CN, which removes almost all of the blue and violet radiation) result in deep red stars that just jump out at the observer. The best example is probably Hind's Crimson Star, a long-period variable like Mira. Just barely making naked-eye brightness over its 432-day variation cycle, the star is more formally known as R Leporis. Though a less colorful example, fifth magnitude 19 Piscium at least shines with a brighter, steadier light.

An aware peruser of tables might have noted three more classes underlying the classic O–M sequence. Newly-discovered, stars of classes L and T are ultracool. Radiating primarily in the infrared, they are so faint visually that there are no naked-eye examples. At the lowest end one can't see such stars at all. Astronomers found them, and in abundance, only when infrared technology was good enough to reveal their long-wave glow. Class L is known not only for molecular oxides (vanadium oxide is especially prominent), but also for strong (to say the least) absorptions from the alkalai metals (lithium, potassium, sodium, and oddly enough, rubidium) and molecular hydrides. In cooler class T we see strong methane bands. Still rather empty, class Y has ammonia absorptions. Cooler than mid-L, the temperatures are low enough to form solid grains, resulting in floating metallic "clouds." Profoundly important, the three classes bridge the gap between stars and planets. We'll meet up with them again later.

The big tool

The magnitude of a star as we see it, the apparent visual magnitude, depends on the star's actual luminosity in the visual band (in watts) and on its distance (specifically on the inverse square of it). Place a star twice as far away, and it's 4 times fainter, 10 times farther and it's 100 times less bright, as first noted in Chapter 4. By the early nineteenth century astronomers had enough stellar distances to enable the calculation of actual luminosities. Stellar distances are principally determined via parallax, the tiny shifts in stellar positions caused by the Earth orbiting the Sun (introduced in Chapter 4). Using satellites, we've measured the parallaxes of hundreds of thousands of stars to more than a thousand light years away. Other methods calibrated by parallaxes take us to the depths of the Universe.

As a substitute for wattage, astronomers use absolute magnitude, the apparent magnitude the star would have at a distance of 10 parsecs, 32.616 light years. At that distance the radius of the Earth's orbit (the AU) would appear a tenth of a second of arc across (see Chapter 4). By default, the apparent and absolute magnitudes are visual magnitudes, those as viewed by the human eye. Other magnitude schemes that use different wavelengths or even total luminosities integrated across the spectrum (bolometric magnitudes) are common, and are used to determine a star's astronomical color and thus temperature, while infrared magnitudes are required for class T and Y stars as they are all so cool that they radiate no visual light at all.

We now have two stellar parameters: spectral class and absolute visual magnitude. With the exception of compositional variants like carbon stars, class is responsive to temperature, and absolute visual magnitude is responsive to actual luminosity or wattage within the visual band. The Stefan–Boltzmann law says that the luminosity of a blackbody is proportional to the fourth power of the temperature; double T and 16 times more energy (2^4) comes tumbling out. All things being equal (which they are not), we thus might at least expect hotter stars to be the brighter, even within the restricted visual band. Around 1915, Princeton's Henry Norris Russell (1877–1957) and Denmark's Ejnar Hertzsprung (1873–1967) independently found just that. The Hertzsprung–Russell (HR) diagram traditionally plots absolute visual magnitude (with numbers getting smaller as we go up to brighter stars on the vertical axis) against spectral class, OBAFGKM, which reflects temperature cooling to the right on the horizontal axis. Starting in the lower right corner of the diagram, as temperature increases toward class B, stars indeed brighten as they climb toward the upper left.

Case closed? In one of the great surprises of the twentieth century, starting in the middle of the diagram there is a branch in which stars perversely become brighter as they cool. The only way this

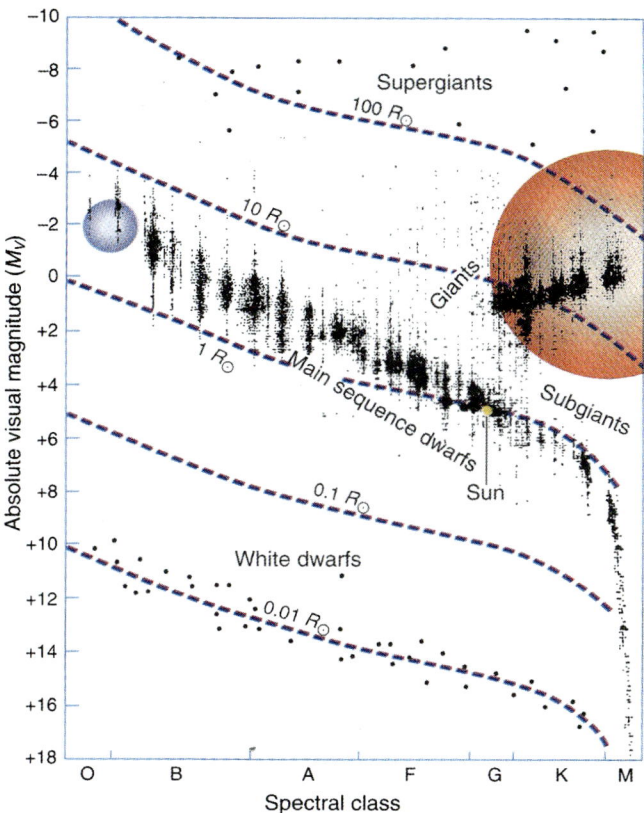

Figure 6.2. The classic Hertzsprung–Russell (HR) diagram, which plots absolute visual magnitude against spectral class, allows us to isolate the various luminosity classes. Dwarf, or main sequence, stars dominate, running from faint red dwarfs at lower right to brilliant O and B dwarfs at upper left. Toward upper right we find first the giants and then at the extreme, the much larger supergiants. The bubbles give a sense of relative size. Subgiants fit between the giants and dwarfs. Running across the bottom of the diagram is a sequence of Earth-size white dwarfs. Selection effects aside, the vast majority of stars are actually red dwarfs, the numbers of stars dropping precipitously as we climb the main sequence, which is a mass sequence that runs from just under 10 percent of a solar mass to more than 100 Sun's at the top. From "Astronomy! A Brief Edition," J. B. Kaler, Addison-Wesley, 1997.

behavior is possible is that the stars in this branch must be progressively bigger, that is, have more radiating surface areas to fend off the lower temperatures (the area of a sphere proportional to radius squared). So, probably without thinking too much about the implications, Russell divided stars into two fundamental kinds that reflect size: "dwarfs," which occupy the main band from upper left to lower right), and "giants," which climb upward from the middle of the band of dwarfs, with (at the time) nothing in between, beginning a charming though perhaps confusing nomenclature still in use. As big as it is, almost a million miles across, the Sun is a class G2 dwarf, Arcturus and Aldebaran class K giants. While so-called dwarfs can be large (10 or more times the size of the Sun) at the upper end of the sequence, giants are much bigger, 10 to 100 or so times the solar diameter. Russell erroneously thought that he had discovered how stars age. They begin as giants and then logically shrink as dwarfs. He was completely wrong. But because of that speculation, cool stars are still called "late" and hot stars "early." At G2, the Sun is an early G star, while at A8, Altair is a late A star. You get used to it.

But that is hardly all. Across the top of the diagram Hertzsprung found a few immensely brighter cool stars that must be bigger than the giants. Here there be monsters, the supergiants of Chapter 4. Of Solar System proportions, the largest (Mu Cephei, also called Herschel's Garnet Star, and VV Cephei) approaches the orbit of Saturn in size.

Then way down at the bottom is a line of anomalously faint stars that run from O to M. The first found and best known is the orbiting companion to the class A dwarf Sirius. Introduced in Chapter 4, it was discovered by Bessel in 1844 through its gravitational influence on the bright star, which was seen to wobble back and forth. Eighth magnitude and only a few seconds of arc from bright "Sirius A" (the brightest star in a double is "A," next usually "B," etc., the letters here having nothing to do with spectral class or with Fraunhofer lines or with anything else), it was not actually seen until 1862 when it was discovered by the American telescope maker Alvan Clark (1804–1887). With broad hydrogen absorption lines, the companion (Sirius B) seemed to be a class A star whose temperature is similar to its brilliant neighbor. However, it's 10,000 or so times fainter, and must thus be tiny, 100 times smaller in radius (square root of 10,000), or about the size of Earth, and correspondingly very dense (typically a million times the density of water). The companion to Procyon and one of the members of the triple system 40 Eridani are similar. Consistent with their temperatures, the three dim companions are white in color. Because the term "dwarf" had already been used for what is now also referred to more appropriately as the main sequence, these dim bulbs were therefore called "white dwarfs." Since the modern collection of such faint stars spans the whole range of temperature, there are blue white dwarfs and red white dwarfs. Sirius A is a dwarf that is white, Sirius B a white dwarf. And they are not the same thing! Astro-nomenclature, historically derived, is strange indeed, but is so ingrained from more than a century of use that it's pretty much unchangeable. Just think of "white dwarf" as all one word and ignore the reference to color.

We are still not done. Much later, Phillip Keenan (1908–2000) of Ohio State University found a smattering of cooler stars even brighter than supergiants, and coined the term "hypergiants," of which there are very few (Eta Carinae of Chapter 4 a prime example). Then there is also a scattering of stars between the giants and the dwarfs. Are they to be "superdwarfs?" "Subgiants?" The latter won the prize. Did we also mention the oddly dim "subdwarfs" that lie to the left of the ordinary dwarfs? Best not, at least not until later.

Within a given spectral class we will find dwarfs, giants, and supergiants. They differ in size and thus in density, that is, in how tightly the various atoms and ions are packed together in the spectrum-forming stellar atmospheres. The change in density alters the spectrum. As an example, the hydrogen lines in class A dwarfs are strikingly wide, while those in class A supergiants are narrow slits, the result of fewer atomic collisions that jiggle energy levels around. Giants fall in between. Among the K stars, giants have weaker cyanogen (CN) bands than dwarfs of comparable temperature class. Once the various rules are calibrated with actual distances (a non-trivial task) we can get the absolute magnitude of a star from the spectrum alone. Comparison with apparent magnitude then gives the distance! With large uncertainties to be sure, but it's probably better than nothing. Among the bigger flies in the ointment is dimming by interstellar dust, which can make a star appear too red and too far away and must be corrected for: see the previous chapter.

In 1943, William Morgan (1906–1994) of the University of Chicago's Yerkes Observatory and Phil Keenan revised the classification scheme by dropping the archaic terms and assigning Roman numerals to the various luminosity classes. Supergiants are class I, giants III, regular dwarfs of the main sequence V. Class I stars are divided into Ia for the brighter supergiants, Ib for the fainter. "Bright giants" between the ordinary giants and the supergiants are Roman II, whilst subgiants are IV. The Sun is thus a G2 V star, Sirius A1 V, Aldebaran K5 III, Betelgeuse M1.5 Ia, Deneb A1 Ia. The concept involves much more than merely a bunch of names and numbers. Morgan, Keenan, and Edith Kellman (1911–2007) went on to establish a grid of spectral subclass and luminosity wherein each bin (at least in principle) has a standard reference star. Comparison with the spectra in the MK photographic atlas (published by the University of Chicago Press) and subsequent modernizations allows the rapid typing of most stars. The importance of the system is that stars are classified only by the appearance of their spectra with no reference to physical parameters. Keenan later added Arabic 0 ("zero," there is no Roman numeral for zero) for his hypergiants.

White dwarfs do not fit into this scheme. The first ones found (Sirius B etc.) had powerful hydrogen absorption lines and were quite forgivably thought to be class A stars. A close look at real class A stars like Sirius A and Vega shows a rich spectrum of metals. However, the spectra of the first white dwarfs known display nothing but the Balmer hydrogen lines with no metals at all. Moreover, there is another set of white dwarfs that have nothing but helium lines and look more like B stars but without the metals, or for that matter any hydrogen. The white dwarfs are chemically distinct from ordinary dwarfs and can't be classed on the Harvard-MK system. From their history they are nonetheless still divided into "DA" and "DB" stars, with a handful of other classes tossed in like "DC" for "continuous spectrum" and "DO" for the really hot ones with helium and without hydrogen. It's simpler to separate them as "DA" and "non-DA." They are plotted on a classical HR diagram according to their temperatures.

On a lesser level, the subdwarfs that lie to the left of the standard, heavily-populated main sequence and were once thought to be too dim for their classes, are actually too hot for their luminosities. They are deficient in heavy elements, sometimes by a factor of a hundred or more, which squeezes them down and makes them smaller and hotter (because their gasses are more transparent and are therefore less subject to the outward push of radiation). The low metal content makes them hard to classify. So are a variety of B, A, and F stars in which some elements settle downwards, while others float upwards again thanks to radiation pressure, giving wacky abundances. It's also worth noting that carbon stars are almost all giants. There seems to be no end to the variety (hence the chapter title). What's the physical basis for it all? We find some of the answers in stellar duplicity.

Double down

About half of our galaxy's stars are in double, triple, even sextuple, systems. The first double star found, in 1650 by the Italian astronomer Giovanni Riccioli (1598–1671), was Mizar, the second star in the handle of the Big Dipper (see Chapter 4). More doubles rapidly followed, until we see that the sky is full of them, our single Sun is almost an exception. (Is it indeed single? There have been several searches for a dim red dwarf companion, none obviously ever found or it would probably have its own chapter. By stirring up the Oort Cloud so as to rain comets down upon us, it did though have a name: the dreaded Death Star.) By definition double stars are physically bound and orbit one another under their mutual gravitational attraction, as first witnessed by (surprise) William Herschel. There are whole hierarchies of them. If the two are well separated, one may be double, making a triple star with the outer orbiting the inner two. The only rule is that the outer star must be far enough away from the inner pair so as not to disrupt their mutual orbit and tear

Figure 6.3.   Most stars are in binary, triple, or even higher-order systems. Here, Theta Serpentis is split into nearly identical class A5 dwarfs 22 seconds of arc (around 1000 Astronomical Units) apart. The spikes on the stars are caused by the interference of light waves as they bend around the supporting struts of the telescope's secondary mirror.   Steve Lewis.

them asunder. The closest star, Alpha Centauri, falls into this category. Closely-spaced G2 and K1 dwarfs (Alpha Cen A and B) orbit each other every 84 years while the pair is rounded by a dim, eleventh magnitude red dwarf (Alpha Cen C, M5.5 V) two degrees, 10,000 AU, away with a period measured in millions of years. (Do the Centaurians know it's there? For that matter, ARE there any Centaurians?) Proxima Centauri (its proper name) is now on the sunside of its path and is therefore actually the nearest star to us, 4.4 light years away. It's really rather amazing that over billions of years Proxima is still attached, unaffected by other stars. If nothing else, it shows how isolated the local stars really are and how rarely they encounter one another.

Continuing on, both stars of a double may also be binary, yielding a quartet, a "double–double" like Epsilon Lyrae in which two close pairs (all four class A dwarfs) orbit each other at a much greater distance. An even more distant one might orbit the inner quartet, yielding a quintuple; while if that one is itself double, we see a sextuple, good examples are Castor in Gemini and the

Mizar–Alcor combination. Mizar's "double" is actually a quadruple (each of Riccioli's pairs themselves double), the quartet orbited by also-double Alcor (A5 V plus a class M dwarf). Castor is made of two bright class A stars averaging about 100 AU apart, each having a close late K or early M dwarf companion; the quartet orbited by a double M1 dwarf. And there it stops. There are no septuple stars, as the outer orbits would just be too large and thus too unstable against stellar encounters. At this point we get non-hierarchical systems like Theta-1 Orionis in the Orion Nebula, whose stars are so mixed together that one or more would (indeed will) eventually be ejected.

There are traditionally three classes of doubles, though the boundaries among them are now quite blurred. "Visual doubles" are just that, seen as binary through the telescope. They range from identical pairs (class F Gamma Virginis, class B Alpha Crucis) to those that differ by 10 magnitudes or more, as in the Sirius system. Some are barely splittable as seen through the telescope, while others are separated by many minutes of arc and thousands of Astronomical Units. Proximity effects give close doubles enhanced colors, making for delightful celestial jewels (as in orange and blue Albireo). The illusion is so striking that astronomers once thought that binary members were physically different from single stars. If the two are close enough together, they can be seen to orbit one another, allowing measure of the orbital period (Herschel again). Mathematically assuming that the lesser orbits the greater, we get the semimajor axis of the secondary star's orbital ellipse (after correction for orbital orientation) in angular measure, and if the distance is known, in Astronomical Units. Application of Kepler's third law yields the sum of the masses (Chapter 2). In reality, the two orbit a common center of mass, which if found (and it's tough to do) gives the ratio of masses, hence the individual masses; stellar mass by far the most important of all stellar parameters.

Sirius is the classic case, the story expanded here a bit from its earlier mention in Chapter 4. Stars orbit the center of the galaxy all on slightly different paths. They therefore move relative to one another. Astronomers break a star's observed motion into two perpendicular parts: the "radial velocity" along the line of sight as determined from the Doppler Effect (described below) and the "proper motion," the shift across the line of sight in seconds of arc per year. The champ of proper motion is Barnard's Star, an M4 dwarf 6.0 light years away in Ophiuchus that clips along at 10 seconds of arc per year. The distance yields the perpendicular velocity in kilometers per second, which when combined with the radial velocity gets us the "space velocity" relative to the Sun. Space velocities are typically 15–20 km/s, although they can range into the hundreds for stars that reside in the galaxy's halo and even higher for those kicked out of the galaxy by its central supermassive black hole. Analysis of local motions shows that the Sun is aiming toward the Lyra–Hercules border at 15–20 km/s depending on the sample.

The Doppler Effect is a change in the frequency or wavelength of any kind of wave as a result of radial, back-and-forth motion of the source or observer. As you row into an ocean's waves, the separation between them seems shorter, in the reverse direction longer. The Doppler Effect is most commonly appreciated via sound waves. The pitch of a speeding car, airplane, or anything else approaching you has a higher pitch than when it passes and then moves away. It's somewhat less appreciated when encountered via a police officer's radar gun, which uses radio waves that bounce off your car. The shift in wavelength is directly proportional to the speed along the line of sight. In stellar astronomy, we use the changes of wavelength of the absorption or emission lines. Great precision in measuring the shifts allows the discovery of orbiting planets.

The white dwarf Sirius B is massive enough to have first announced itself in 1844 by wobbling the proper motion path of vastly brighter Sirius A. After its sighting in 1862, long-term observation of the 50-year orbit about the center of mass (the two averaging 20 AU apart) revealed that Sirius A, a rather ordinary class A dwarf, carries 2.1 solar masses. The surprise is that tiny Sirius B, with a luminosity 2.4 percent that of the Sun, a temperature of 24,800 Kelvin, and thus a radius 92 percent that of Earth, weighs in at a full solar mass. The average density must therefore approach an astounding two metric tons per cubic centimeter (as already noted in Chapter 4). Auguring the next chapter, we are looking at the old nuclear-burning core of what was once a mighty star that dominated our present Sirius A.

The members of a binary can be so close that they are unresolvable, traditionally under half a second of arc or so. But in that case, their orbital speeds must be high. Their spectra would then be subject to significant periodic Doppler shifts that move the absorption lines back and forth to a degree proportional to the stars' velocities, which yields the orbital periods and thus the mass ratios. But since we do not necessarily know the orbital tilts, we can get only lower limits to the individual masses. While the result may seem limited, it's statistically important and as we will see of immense significance in the study of other planets orbiting distant stars. Modern optical interferometry, however, can break at least part of the barrier by observing stellar members that are remarkably close, spectroscopic binaries becoming visual doubles but with the mass ratio known from the velocities.

However, if the tilt of the orbital plane is such that it's sufficiently close to the line of sight, then the stars might eclipse one another twice each orbital period and we have hit the mother lode of binary studies. Given velocities, we can obtain a huge amount of information that includes stellar masses, temperatures, surface brightnesses, even shapes. Chief among eclipsers is Algol, Beta Persei, the "Demon Star," which drops by 1.2 magnitudes every 2.87 days as a K giant passes partially in front of a much smaller B dwarf. The secondary eclipse that occurs when the B dwarf takes a bite out of the K giant is not noticeable to the naked eye. Beta Lyrae is another good one. The members of both it and Algol are so close as to be mass-transfer binaries in which tidal distortions force one star to pass matter to the other via an accretion disk. In Algol's case, the B dwarf is consuming the K giant, and it will probably continue doing so until there is nothing left but the giant's old nuclear-fusing core, which will have become a dense white dwarf. The members of Beta Lyrae are in near contact with each other. Some eclipsers are even odder. Among the weirdest is Epsilon Aurigae, wherein every 27.1 years, a class F supergiant is eclipsed by a huge dust-shrouded star or stars for two years. The supply of oddities seems endless. Another reason for the title of this chapter.

O'er the bounding main

Decades of binary-star studies have slowly accumulated precise measures of the masses of all sorts of stars. Of first importance, the main sequence, the set of dwarfs from class M to O, is a mass sequence. Its lower limit is theoretically defined by the cessation of full nuclear fusion within the stellar core, in which hydrogen nuclei are turned into helium, specifically He-4. Below around 8 percent the solar mass, the internal gravitational compression caused by the star's outer layers is no longer sufficient to raise the temperature high enough to run the process. Another way of saying it is that fusion can no longer support the star, and it is fated to steadily contract, radiating away its gravitational energy until a phenomenon called "degeneracy" (critical to white dwarfs and described in the next chapter) stops it.

Climbing upward along the main sequence, the stellar luminosity increases at a great rate. The formal, averaged, mass–luminosity relation states that the luminosity, the total energy output over all wavelengths, goes up as the mass (M) raised to the 3.5 power, that is between M cubed and M to the fourth power (M squared-squared), the immediate exponent varying some with the mass range being examined. But basically, if you double the mass, the total brightness of the star increases by a factor of between 8 and 16. Starting near 0.08 solar masses where class M runs into class L, we climb through 0.8 solar masses at K0, 1.0 at G2 (the Sun), 1.5 at F0, 2.5 at A0, then a huge jump to 20 Suns at B0. See the summary table, which for various spectral classes gives the mass, surface temperature, luminosity, and hydrogen-fusing lifetime, which rapidly goes down as the mass and fuel supply go up, a critical matter deferred until the next chapter.

The numbers in the table are all rough averages. Stars slowly brighten as they use their internal hydrogen fuel supplies. At the beginning of their hydrogen-fusing lives they are significantly dimmer, while at the end, brighter and cooler. The early Sun was 30 percent fainter when it was born 4.6 billion years ago and at the end of its main-sequence life will be twice as bright as it is now, which rather obviously has profound implications for life on our planet.

Table 6.2. Main sequence mass–luminosity–age relations.[a]

Class	Mass (Suns)	Temperature (K)	Luminosity (Suns)	Life (yrs)
O2	120?	52,000	2 million	1 million
O5	40	44,000	1 million	5 million
B0	20	31,000	30,000	80 million
B5	5	15,000	1000	95 million
A0	2.5	10,000	100	600 million
F0	1.3	9500	30	5 billion
G0	1.1	6000	10	7 billion
K0	0.8	5300	3	16 billion
M0	0.5[b]	3700	0.1	
M5	0.25[b]	3000	0.02	

[a]Approximate average.
[b]Lifetime greater than that of the galaxy.

We left off at the high end of the B stars, where we run into the amazing O stars at the top of the sequence. They, along with the hottest (earliest) B stars (B0 to B1) are different. Masses scream upward, the luminosities increasing ever faster, with temperatures high enough to ionize the surrounding interstellar clouds (if there are any around) to produce diffuse nebulae, as does Theta-1 Orionis C, which gives us the gift of the Orion Nebula. Though we know the mass at the bottom of the main sequence, we really have little idea of what lies at the top: 120 Suns at class O2 is commonly assumed, but it may be much higher. The mass–luminosity relation does not include giants, supergiants, or anything else. It belongs strictly to the common main sequence. For the other kinds of stars we have to look to stellar evolution.

What the lower main sequence lacks in individual masses, it makes up for in numbers. Around 70 percent of the stars in the Milky Way belong to class M (leaving out populous L and T). They are everywhere. Yet M dwarfs are so dim that not one is visible to the naked eye. The coolest dwarf that could by itself be seen without optical aid is around class K7, represented by 61 Cygni B, which with 61 Cygni A was (because of its high proper motion) the first parallax star. G stars like the Sun account for only seven percent of dwarfs, and then the bottom drops out: one percent for class A, 0.1 for B, 0.0004 percent for class O! The scarcity of O stars is a good thing since O stars eventually explode and you would not want to be close to one. Since they are rare the odds are that the closest will be far away, as they are today, but not necessarily were in the distant past.

Here we encounter strong observational selection, which is a plague unto astronomers. Nature lets us see what she wants. Most of the constellations are made from A and B dwarfs and K giants. While not all that common per million cubic light years, they are bright and easily seen, so they dominate the sky. Remove the M dwarfs and you'd not notice a thing. Take out the A dwarfs and there goes the Big Dipper, knock off the B dwarfs and Scorpius would go back into hiding. Class O dwarfs are so rare that there are few to be seen without the telescope, though if we were to eliminate them part of Orion disappears, including the brightest star of his head and the Orion Nebula, not to mention almost all of the diffuse nebulae of the Milky Way.

Gatherings

Stars not only like to double up etc., they like to congregate, to cluster, which is a clue to how they are born. The Milky Way is rife with loosely organized open clusters like the Hyades, the Pleiades, and the Beehive. Some, like Messier 11, are extraordinarily compact, dense with stars, while others — Coma Berenices comes to mind — look like they are falling apart. Which they are. The gravitational bonds that keep them together are relatively weak, so stars gradually leak away. After a few billon years not many stars are left to mark what may have been a mighty assembly. The Sun was probably born in a cluster, but left early on, its mates now nowhere to be found (though we have looked for them). A cluster's viability depends a lot on its birthplace and surroundings. They don't last long in the galaxy's interior because of tides raised by nearby stars. In the sparse outer galactic disk they can last practically forever.

The gigantic poorly-populated halo that encloses our galaxy's thin disk is home to the much more compact, tightly bound and ancient globular clusters like Messier 13, 47 Tucanae, and a variety of others (Chapter 4). The densest have probably survived intact. Some are so compact that black holes might grow at their centers. With the application of the theory of stellar evolution, clusters give us a timeline, a way of looking back into the past, telling us that the Milky Way is at least 12 billion years old. It was probably born in lesser form right after the creation event, the Big Bang. Some of the galaxy's stars can take us most if not all of the way there.

7

Star Lives

While there are those who reject anthropomorphism in science, it remains a metaphoric, poetic way of expression that infuses astronomy. We speak of stars being born, of passing through youth, middle age, old age, of dying, as if they were sentient beings. Thus W. H. Auden's verse aside ("you can look it up"), in our hearts they can be friends that are with us so long as we, along with Whitman, walk the Earth.

Growing up

Released from their placental dark clouds, stars begin their lives with fixed fuel supplies and, like ourselves, like anything, perhaps even the Universe itself, cannot last forever. Eventually the fuel runs out. Might some of the "infinite variety" of the last chapter be explained by mass combined with age? Observation and theory show us that main sequence dwarfs are remarkably stable affairs, demonstrated not only by the long term steadiness of the Sun as revealed through the Earth's fossil and geological record, but also by the main sequence's very existence, with its huge numbers of stars stuffed into a narrow band in a plot of luminosity vs. temperature or spectral class (Figure 6.2).

Stars can shine under their own gravitational compression. After all, that's how they get to be stars in the first place (Chapter 5). Running off "Kelvin–Helmholtz contraction" (after Lord Kelvin, introduced in Chapter 1, and the German physicist Hermann von

Helmholtz, 1821–1894), a star like the Sun can release enough gravitational energy by shrinking (falling into itself) to allow it to operate at its present luminosity. We can even reproduce the mass–luminosity relation. At the necessary contraction rate of roughly 10 meters per year, however, the Sun just cannot shine that way for long, far less time than required by the fossil record. Dwarf stars, those of the main sequence, are instead supported by hydrogen fusion, which supplies their energy so that gravity does not have to do the job.

In any stable star, the inward pull of gravity is countered by the outward push of gas pressure. The pressure of a gas within a specified volume depends directly on the temperature and the number of particles. Critically, pressure does not depend on the KINDS of particles, whether big ones or little, just their number. The rate of solar fusion is highly sensitive to temperature, proportional to around T to the fourth power. Double the temperature and the burning rate (nuclear burning, fusion) increases 16 times. As hydrogen is converted to helium within the core, the number of particles goes down. But the outward push must be maintained, so under the influence of gravity, the core shrinks, which results in an increase of both density and temperature and which in turn makes the reactions run faster. The balance between fuel supply and reaction rate is almost perfect, allowing main sequence stars to achieve their great stability over long periods of time, indeed, allowing a main sequence at all. Our very lives are dependent on this extraordinary stability.

But there is that sneaky "almost" to deal with, as the fusion rate slightly overcompensates. Unlike a wood fire, as its fuel supply diminishes, the star does the opposite of that expected, becoming brighter rather than dimmer. At the same time the star as a whole slightly expands, which causes the temperature to change at the surface, increasing or decreasing depending on circumstances. As noted in Chapter 6, when the Sun was born 4.5 billion years ago it was 30 percent fainter than it is now, which brings up a curious problem, as the Earth should have been much colder, too cold to sustain life (as if we know). Yet life came along very quickly, within a billion years. While there is no definitive solution to the "faint young Sun paradox," two possibilities come to mind. The solar wind drags the Sun's magnetic field out with it. Still attached to the Sun, the field lines act like ropes that slow the rotation. With the Sun spinning much faster in its early days, solar activity and its potential heating effects (of which we are quite ignorant) could have been much greater. Alternatively, the carbon dioxide level could have been higher, raising the greenhouse effect and consequently the terrestrial temperature. Or for that matter, both. Or something else. We don't know.

Meanwhile, back in the present, the Sun has another five billion years of core hydrogen fusion, and thus main sequence life, left to it. But the continued slow brightening will probably make life impossible a billion years hence. Again like we actually know: look at the places where we find life now. But with the water all boiled away, it seems reasonable to assume that not much life, if any, will survive. Can we go some place else, to some other youthful star with a long life ahead of it? Quite possibly: at least we might send a representative sample and start all over again. Five or so billion years from now the Sun will be twice as luminous as it is today as it prepares for its glorious death. Such change has nothing to do with global warming, as it is far too slow. On the average, the Sun increases its luminosity by 0.000000004 of a percent per year, a hopelessly undetectable amount. The climate-change deniers will have to find something else.

As we go upward in mass along the main sequence, the fusion mechanism changes. The Sun runs largely off the proton–proton chain. But not entirely. Seven percent of solar energy comes from the "carbon cycle." On Earth the term refers to the way in which carbon is cycled through life, rocks, oceans, and the atmosphere. In stars it's another way of turning hydrogen into helium, with the carbon atom acting as a nuclear catalyst. Taking six basic steps, the cycle's just a bit more complex than the three-step p–p chain. It starts when a carbon-12 nucleus captures a proton to make the next heavier element, nitrogen, in the form of the isotope N-13, the process releasing a heating gamma ray. But N-13 is highly radioactive (unstable against decay) and in step two it quickly ejects a positron (a positive electron) and one of those ubiquitous neutrinos (Chapter 1). With one of its positive charges released the N-13 turns into C-13. While the neutrino goes scooting off, the positron doesn't get very far before it mutually annihilates with an electron, kicking off two gammas. Pretty common, stable C-13 makes up about one percent of the carbon within your body, or within just about anything on Earth for that matter. The rest is carbon-12.

Not content to just sit there, the C-13 nucleus catches another proton (number 2) to make familiar nitrogen-14 (which you are breathing) and yet another gamma ray. Continuing on, in step four, N-14 grabs a third proton to become oxygen-15 (and, as now expected, a gamma ray). The radioactively unstable O-15 decays back to stable N-15, a positron and neutrino fleeing the scene with the usual results. To end the process, N-15 picks up a fourth proton. One would think from the pattern above that oxygen-16 would come out of it. Instead, somewhat more than 99.9 percent of time the whole thing falls back to carbon-12 with the release of a helium-4 nucleus. And there we finish, with four hydrogen atoms (add them up) turned into one atom of helium. The result is the same as the p–p chain, with the original carbon ready to do it all over again. "But wait, there's more." Just under 0.1 percent of the time, the last reaction indeed does go to oxygen-16 and yet a further set of reactions, which gives the whole process the more accurate name, the "CNO cycle." Note that not only is the star producing energy, but also heavier and heavier elements.

While the p–p reaction chain is highly sensitive to temperature, it's nothing compared with the carbon cycle, whose rate goes as the 15th power of the temperature (or more: it's not fixed). So as we go to higher stellar masses and the core temperature climbs, the carbon cycle rapidly becomes ever more important, equaling the energy output of the p–p chain around class F5 at 1.25 solar masses. With only a bit of an increase in mass and core temperature, the carbon cycle takes over entirely, while the contribution of the p–p chain becomes inconsequential.

When stars are born, assuming more or less similar (usually solar) chemical abundances, they fall along a pretty narrow line in the HR diagram (converted perhaps into the theoretician's diagram, which plots the logarithm of luminosity against log T) called the Zero Age Main Sequence (the ZAMS). The slow changes in surface temperatures and luminosities produced by the decrease in the hydrogen fuel supplies act to broaden the main sequence from a line into a band with the ZAMS as the defining left edge. Further broadening is produced by lowered metal abundances as found in the subdwarfs, which shift the ZAMS and the whole main sequence in the other direction, a bit toward higher temperatures.

As birth mass increases, so does the size and mass of the core and with it the fuel supply. But given the huge dependence of burning rate on core temperature, the speed at which the core consumes its hydrogen increases precipitously, as does the luminosity, theory nicely matching observation and the empirical mass–luminosity relation. The result is that higher mass stars live dramatically shorter lives than lower mass stars. Our anchor, the Sun, has a core-hydrogen-fusion lifetime of 10 billion years. At three solar masses, the ZAMS luminosity is more than 100 times that of the Sun, while the lifetime decreases to 300 million years. At 40 Suns, L(ZAMS) hits a quarter million solar luminosities, and the lifetime has dropped to five million years. Class O dwarfs are literally here today gone tomorrow, one of the reasons for their rarity. Their short lifetimes don't allow most O stars to move very far from their birthplaces before they die, giving rise to "OB associations," expanding clumps of massive (and also lower mass) stars not bound together by gravity, as are clusters. Most of Orion, Scorpius, and Perseus are made of associations of related stars.

We can gauge the age of a dwarf star since birth by its position on the main sequence, by just how much it has changed its luminosity and temperature as compared to the values they had on the ZAMS. In practice, it's a bit tough to do, as the temperature can be difficult to measure and may not even have a unique value, and the luminosity is dependent on distance and estimates of the amounts of ultraviolet and/or infrared radiation. There are other routes to main sequence ages. Lithium is easily destroyed by nuclear reactions that take place at relatively low temperatures. If a star has a convective envelope that takes the photospheric gases to sufficient depth (roughly, cooler than mid-A), we can estimate the star's age by how much lithium remains. At a more or less advanced age, the Sun's is nearly gone. And since magnetically active stars slow with age, rotation speed works too.

At the other end of the scale, toward lower mass, are three important limits. Below 0.85 of a solar mass, around class G8, the dwarf lifetime exceeds that of the galaxy. No star with a mass under that limit has ever died. It's one reason that there are so many red dwarfs (the other being that nature just likes to make them, as opposed to O dwarfs). Then below about 0.08 Suns, the core temperature becomes too low to allow full hydrogen fusion, though the fusion of deuterium (emplaced at birth) into helium can still proceed. These "stars," brown dwarfs of classes L, T, and Y, cool rapidly as they separate themselves from true red dwarfs. They are supported not by fusion but by "electron degeneracy," which is more a subject for white dwarfs and will be dealt with later. Finally, below 1/80 of a solar mass, 13 Jupiters, the internal temperature drops so low that even deuterium fusion ceases, and we supposedly enter the realm of the planets. Planets almost by definition are built from the ground up, by accumulation of circumstellar dust grains as the parent star is being born, while stars (if we extend the original definition of Chapter 4) are created from the top down, by contraction of bloblets in molecular clouds. Are there crossovers? Can some brown dwarfs be made like planets, or planets like stars? Here there be dragons in a terra incognita where little is known and what we think we do know is probably wrong.

And then the hydrogen runs out

And then also the fun starts: at least it's fun if we are watching from another planetary system. When the fuel supply (hydrogen) is gone and energy production from fusion stops, the star's core, now a solid block (sphere actually) of helium-4, loses its support. It then has no choice but to contract much more rapidly, which releases gravitational energy and makes the core heat: again just the opposite of what you might expect when a fire is quenched. The heat spreads outward into the surrounding hydrogen-rich envelope (as it has been doing all along), which causes hydrogen fusion to run within a shell around the now-dead, collapsing helium core. A sunlike star (or one of a few solar masses) now takes a sudden turn to the right on the HR diagram (Figure 6.2), expanding and cooling at its surface. First behaving as an intermediate "subgiant" (see Chapter 6), the star next begins to swell and brighten at a more or less constant temperature into a real red giant, to a thousand times the current solar luminosity. The core, still surrounded by its hydrogen-burning shell, shrinks to near the size of Earth while the outer photosphere expands to the sizes of the orbits of the inner planets. Mercury is doomed. Enjoy it now. If earthly life had not ended at the end of main sequence life, it surely will end now.

Figure 7.1. The future Sun, 1000 times brighter than it is today, its diameter approaching the orbit of Venus, rises over Earth's parched landscape just before it fires up its helium core. From "Stars," J. B. Kaler, Scientific American Library, Freeman, New York, 1992.

Stellar luminosity, *L*, equals the radiant energy per unit surface area times the surface area, which for a sphere is proportional to radius squared. The Stefan–Boltzmann law of Chapter 1 states that the radiant energy per unit area of a blackbody, which the star is assumed to be, is proportional to temperature to the fourth power, so luminosity, *L*, is proportional to radius squared times temperature to the fourth. Increase *R* at constant *L*, and *T* quickly drops, etc. Why do evolving stars behave this way? There seems to be no easy intuitive answer. In the words of a world expert, "that's what the calculations show they do."

Iron, atomic number 26, has the most tightly bound of all atomic nuclei. There is no way of extracting any further nuclear energy from it. In broad terms, you gain energy by fusing lighter atoms together until you get up to iron, or by fissioning (breaking apart) heavy radioactive elements (uranium, thorium, and the like) until you get down to iron. All we've done so far is to fuse hydrogen into helium so there is plenty of room, even considering the manufacture of some nitrogen, carbon, even oxygen isotopes via the CNO cycle. When the temperature of the shrinking core hits 100 million Kelvin, the helium atoms are moving with such vigor that they can overcome their mutual electrical repulsion and start to fuse together, which creates a whole new energy source. But it's not easy. The product of the fusion of two He-4 atoms is beryllium-8. Hopelessly unstable, it immediately decays back to He-4. To make helium fusion work, another He-4 nucleus has to hit the Be-8 within its hurried life of a ten-millionth of a billionth of a second, which results in carbon-12. In effect, helium fusion requires the simultaneous collision of three He-4 atoms. Since helium nuclei were called "alpha rays" before anybody knew what they were, and are still commonly called "alpha particles," helium fusion is run by the "three-alpha" or "triple alpha" process. Since almost all heavy elements come from stars, there are only small amounts of lithium, beryllium, and boron (elements 3, 4, and 5) in the Universe. The trio is skipped over, number 2 helium going directly to number 6 carbon. A hit with a helium nucleus makes oxygen. "Alpha capture" can work its way up even farther, which is why even-numbered elements are more popular than elements with odd numbers of protons.

With a new source of support, the core expands, the outer envelope shrinks, the luminosity declines, and the star (and our future Sun) becomes one of the huge numbers of cool class K orange giants in the sky. Most of the energy, some 90 percent, of complete conversion to iron is in the fusion of hydrogen into helium. Helium fusion has roughly 10 percent the capacity of the total, so the stable red giant phase does not last for long, around 10 percent the time of hydrogen fusion. In a strong sense, the red giant branch of the HR diagram is a temporary main sequence for helium-burners.

The age of a star cluster, whose stars were born all at the same time, can be found from the so-called "giant turnoff," from the most massive stars left on the main sequence, whose core-hydrogen-burning time can be calculated from theory. If, for example, the most massive stars left carry just under a solar mass, the cluster must be about 10 billion years old, the main-sequence lifetime of the Sun. The open clusters of the Milky Way's disk show a huge range, from literally zero (just born, with class O stars intact) to about 10 billion years, which, since we see none older, is pegged as the age of the disk itself. The globular clusters of the halo, however, are older, in the neighborhood of 12 to 13 billion years, the main sequence (corrected for their lower metal abundances) stopping around 0.8 solar masses. It's then obvious that the halo came first, the galaxy later collapsing to form the disk, with mergers with other smaller galaxies playing an equally strong role.

Then the helium runs out

Do we see a pattern here? Eventually, the helium fuel in the core is exhausted, all of it turned into carbon and oxygen. Reverting to its earlier behavior, the core once again contracts and the star brightens even more than it did before as helium fusion moves out into a surrounding shell. The helium-burning shell is in turn surrounded by the old hydrogen-burning shell. The two turn on and off in sequence, the onset of helium fusion explosive (though the detonation may not directly affect the stellar surface). Gravity again takes the upper hand. The energy released from the second core contraction vastly increases the stellar luminosity, the future Sun now reaching upwards of 6000 times its current brightness. The radius again increases to even larger proportions while the stellar surface cools yet more, the star becoming redder, moving into class M. If Mercury was not destroyed before, it surely will be now. Even Venus seems doomed as the solar radius grows toward the size of Earth's orbit. The graphical path such stars take on the second ascent of the giant branch of the HR diagram is sort of asymptotic to the first path when the stars had a dead helium core. They are thus referred to as "asymptotic giant branch", or "AGB" stars. You can actually subscribe to the "AGB Newsletter" should you so choose. It's free.

Figure 7.2. At the end of helium fusion, the Sun, more than 5000 times its current brightness, will approach Earth's orbit, perhaps destroying our planet, though huge mass loss may send it reeling away. From "Stars," J. B. Kaler, Scientific American Library, Freeman, New York, 1992.

Saved?

This behavior can't go on forever, else the sky would be filled with brilliant dying stars, which it isn't. The end, and perhaps the salvation of Earth, is achieved through stellar winds. Though it hardly seems that way to us when our power goes out during a solar storm (a coronal mass ejection), our Sun pumps out a really weak wind, losing a mere hundredth of a trillionth of itself per year. As giants get ever larger, though, their surface gravities go down, allowing for stronger winds.

The wind's origin changes too. The solar wind is driven by the Sun's magnetism. AGB winds are driven by pulsation and radiation. As they brighten, the advanced giants become highly unstable, varying hugely over periods of months and years. The prime example is Mira, a class M7 or so red giant that has a "now you see it, now you don't" relation with its constellation Cetus (see Chapters 5 and 6), varying from second magnitude to about tenth over roughly a year (actually, 334 days). Large numbers of these Mira variables (or long-period variables, LPVs) populate the sky, many reaching naked-eye brightness. The pulsations, accompanied by emission lines in LPV spectra, push gas outward, where it can condense into dust grains. The fierce radiation then pushes on the grains, which couple to the gas to drive powerful stellar winds that can blow billions of times more strongly than what we experience from the Sun. Increasing in power, the AGB wind becomes so intense that it removes most of the stellar envelope, leaving behind not much more than the old nuclear-burning carbon/oxygen core, which terminates the increase in luminosity at a level set by the original (and final) mass. Thus the future evolving Sun may never grow enough in size to reach us. The mass loss will also surely alter our orbit, perhaps making the Earth spiral outward. It might even be lost to the Solar System. Some Mira variables become completely buried within their own ejecta and are detected only through their strong infrared radiation, most of which comes not from the star itself, but from the heated dust cloud. Molecules form within the escaping mass as well, some radiating as powerful masers (Chapter 5).

At this point, if the star has the right mass, convection currents might sweep some of the internal carbon upward, causing the star to dramatically alter its surface chemical composition, whence it becomes a deep red "carbon star" of the sort known since the mid-nineteenth century. At various other points of evolution, freshly made nitrogen and helium can be dredged up too. The dust in the star's wind is silicate-rich for ordinary advanced (AGB) giants, carbon-rich for carbon stars. The stuff then gets wafted out into the cosmos by the stellar winds. Here is the origin of most of the dust in the Milky Way, the same dust that helps make molecular clouds and form stars, stellar death again seen to perpetuate stellar birth. As important, here is also the origin of almost all the carbon in the Universe, our lives oddly completely dependent on the hosts of stars that passed through the AGB phase before the Sun was born.

Of crucial importance, far down within the helium-burning shell, elements heavier than iron can capture free neutrons, which with the ejection of negative electrons allows manufacture of a variety of heavy elements all the way to bismuth, element number 83. These freshly made nuclei can also get dredged up to the stellar surface. Proof that "slow neutron capture" (the "s-process") operates in these advanced giants is the observation of element 43, technetium. All of its isotopes are so unstable that there is none on Earth save for a tiny bit manufactured for medical use. Technetium's very existence in a star shows that it is being made as we watch. Nearly all our zirconium, and good fractions of other heavy elements, were created by this process and deposited in our birth cloud before the Sun was born. Here is one of the great themes in astronomy, that other than hydrogen and helium and a smattering of lithium (which came out of the Big Bang), our chemical elements were made in (or by) stars and distributed into the cosmos through winds and (as we will see) more importantly through stellar explosions, the new elements eventually to be incorporated into later stellar generations.

The end is nigh

As the wind reveals the dying core, the surface temperature of the star increases while the luminosity stays constant at the value set by the onset of the final strong wind. Now a thinner wind from the stellar cinder blows ever faster, which shovels the interior of the escaping mass in the prior wind into a dense, dusty shell. When the surface temperature of the old nuclear-burning core hits around 30,000 Kelvin, its growing ultraviolet radiation begins to ionize the expanding shell, giving rise to one of the most beautiful of celestial sights, a planetary nebula (see Chapter 4 and Figure 4.6). A misnomer, the nebulae got their generic name from their discoverer, William Herschel, who had found Uranus four years earlier and who meant the term to mean "disk-like."

A typical planetary nebula (if there is such a thing) appears as a disk or spheroid composed of multiple shells and arcs, the structure reflecting the complexity of the wind that made it. At the center is always the star that pumped it out. If the core of the old giant is orbiting within a binary system, the stirring of the wind may make the nebula even more complex. It's possible that stellar duplicity may even be necessary to construct the object in the first place. Will the Sun produce a planetary nebula? We don't really know. And we won't find out either, at least not directly.

As the remainder of the stellar envelope is swept away and the raw core becomes ever-more revealed, the surface temperature rises even higher. Typical temperatures are far greater than those of class O dwarfs, upwards of 100,000 Kelvin. More than 200,000 K is not unknown. With luminosities measured in the thousands, even tens of thousands, of Suns, the central stars of planetary nebulae are among the hottest and most luminous stars in the Universe. But with most of their radiation in the ultraviolet they tend to be visually faint, and through the telescope are usually quite unimpressive.

Favorite targets are the Ring Nebula in Lyra (Messier 57, with its 15th magnitude central star; see Figure 4.7), the Dumbbell Nebula in the modern constellation Vulpecula (Messier 27, beautifully set against the Milky Way), the Cat's Eye in Draco (within which Herschel discovered the first central star and was the first to be imaged by Hubble), the Saturn Nebula (Herschel's discovery object) and the huge Helix (a quarter degree across), both in Aquarius, and the Eskimo in Gemini (which has the visually brightest central star). Their names alone tell of their beloved natures.

Figure 7.3. The central exciting star of the planetary nebula NGC 2440 (see Figure 5.2) is on its way to becoming a white dwarf. With a temperature around 220,000 Kelvin it's one of the hottest stars known. The nebula, formed from the last phases of stellar mass loss and now around 1.5 light years across, is dissipating into interstellar space, where it will deposit a fresh load of carbon previously made in the furnace of the stellar interior. NASA, ESA, and K. Noll (STScI).

Planetaries are also favorites of the Hubble Space Telescope, which has imaged dozens of them. Many, most famously the Ring and the Cat's Eye, are surrounded by huge outer shells that tell of earlier stages of significant mass loss through winds.

The emission line spectra of planetary nebulae show that a good fraction of them are enriched in nitrogen, helium, and carbon, and presumably also in heavier elements that include those made by slow neutron capture. The nebulae are carrying away the chemical elements that were made in the stars during the red giant branch stages before the nebulae were born from the giants' ejected winds. We are seeing the chemical enrichment of interstellar space, and quite clearly carry within us the by-products of stars that lived and died before the Sun was born.

At some point that depends on the mass of the core, hence on the mass of the original star, the core loses practically all its outer skin, stops heating, and begins to cool and dim. At the same time the surrounding nebula, which started life as a mere dot, has expanded to more than a light year across and will in a few tens of thousands of years begin to meld with the cosmic interstellar gases, leaving the core behind. At first shrinking, the core becomes denser and denser as it turns into a stable white dwarf, the surrounding nebula disappearing altogether into the cosmic gloom. This process cannot go on forever either, else the white dwarfs would contract to nothing and instead of the myriads that surround us we would hardly see any of them. Something must keep them from dying altogether.

Degeneracy

Photons behave both as waves and as particles. So do electrons and other subatomic particles. The wave character means you can't really pin electrons (or for that matter, any other particles) down. Instead of "orbiting" their nuclei, electrons more surround it in a wavelike fashion. The high internal temperatures of the white dwarfs cause the atoms to become highly ionized, which frees the electrons into a gas of their own. As the density approaches a ton per cubic centimeter, the wave character of the electrons causes them to interfere with one another to the point where they can get no closer together, and they are then said to be "degenerate." Collectively halting the contraction, they permanently stabilize the white dwarf, which has no future but to cool forever. The same phenomenon also supports brown dwarfs. The cooling time for white dwarfs is so very long, longer than the age of the galaxy, that every white dwarf ever made is still visible. There is no such thing as a "black dwarf," a prior white dwarf that has cooled to invisibility ("black holes" quite different beasts). There are thus a lot of white dwarfs to study.

Such will be the fate of the Sun, which is expected to turn into a white dwarf with just over half its current mass. A fundamental fact of stellar life, one that cannot be stressed strongly enough, is

Figure 7.4. The class A1 dwarf Sirius A (which appears as a disk because of overflowing pixels; it's really a point) overwhelms its white dwarf companion, which is seen as a blip down and to the left, the little star smaller than Earth but a million times denser. All stars born with less than 10 or so solar masses end up this way, destined to cool nearly forever. NASA, H. E. Bond and E. Nelan (STScI), M. Barstow and M. Burleigh (U. of Leicester, UK), and J. B. Holberg (U. of Arizona).

that stars die with much less mass than they were born with. The remainder is recycled into new stars. Most white dwarfs, the DA variety that includes Sirius B and Procyon B, are left with thin hydrogen layers that give them the superficial appearance of class A stars (see Chapter 6). Others, the "DB variety," have lost this layer and show helium absorptions instead. Unfortunately it's not that simple (what is?) as they seem to be able to turn themselves from one kind to another.

Perhaps the most important feature of the white dwarf state was worked out early in the twentieth century by a young Indian astrophysicist, Subrahmanyan Chandrasekhar, while on his way to England to work with Sir Arthur Eddington, one of the great figures of science.

If you increase the mass of a white dwarf, the internal temperature goes up. The higher the temperature, the faster the electrons move. Inevitably, their speeds as they interact approach that of light, and the rules switch over to Einstein's relativity. Chandrasekhar found that when a white dwarf hits a critical mass of 1.4 times that of the Sun, the degenerate electrons can no longer provide support, and it has to collapse. White dwarf masses are closely tied to the mass of the original star. The Chandrasekhar limit is breached by stars with birth masses greater than 8 to 10 Suns. Beyond that vaguely-known limit, stars can no longer make white dwarfs. Instead they make something else, something far more spectacular.

Supernovae

Nearly 1000 years ago, in the year 1054, humanity was treated to a spectacular outburst in Taurus near the beast's southern horn. Those ardent recorders of celestial events, Chinese astrologers and government officials, described its behavior and position quite well. Though nearly as bright as Venus at her best, little mention was made of it in Europe. Not only was it dangerous to go outside, it was even more dangerous to comment that the skies were not changeless: as Galileo found out. At the site of the brilliant "new star" (or so it then seemed) we now find the equally spectacular Crab Nebula (Figure 4.11). More than a tenth of a degree across, it's expanding at 1500 kilometers per second, which places its origin right back to around 1054. We now know it to have been a supernova, an exploding star.

Stars that make white dwarfs, in the range from just under a solar mass to 8–10 solar, cannot get hot enough in their cores to fuse carbon and oxygen much farther up the element chain. At the very top, the C/O mix might "burn" to oxygen and neon, giving us rare neon white dwarfs. Stars born above 8–10 solar masses, however, those of class O and a few of the hottest of B, are so massive and get so hot inside through compression that they can run the fusion chain to much heavier elements. When the core hydrogen runs out, they first swell with dead helium cores to become supergiants, which (like ordinary giants) are stabilized by the fusion in the core of helium into carbon and oxygen. At the same time, hydrogen is burning to helium in a shell around the core. In the lower mass range they hang out as brilliant class M red supergiants like Betelgeuse and Antares. Above about 40 to 60 solar masses they may return to, or get stuck in, the blue high temperature portion of the HR diagram.

When the helium runs out, the C/O core is no longer destined to become just another white dwarf, but can easily contract until it gets hot enough to fuse to a mix of neon, magnesium, and oxygen, as above. Helium fusion now moves out into a shell that is surrounded by a larger hydrogen-burning shell. And so it continues. Once this more advanced nuclear burning is completed, and the core has nearly pure mix of the three elements, it again contracts and heats until it fires up the Ne/Mg/O into a mixture of silicon and sulfur, the previous stages of fusion wrapped around the core like an onion.

The eventual Si/S core then begins to fuse at last to iron. Each fusion stage provides less energy as iron is approached. But each stage must support roughly the same load, the weight of the surrounding star. As a result, each burning phase just lasts a shorter period of time. Hydrogen fusion to helium might take millions of years, helium fusion perhaps a tenth of that, whereas silicon burning may last just a couple of weeks. But as we saw above, iron can't fuse with itself or with anything else to produce energy, as the nuclear energy is nearly all extracted. When the iron core, basically an iron white dwarf near the Chandrasekhar limit, contracts there are no further fusion stages left to stop it. The core can no longer provide support, and it undergoes a catastrophic collapse with a speed a good fraction that of light. The density becomes so great that the iron atoms break down and all the freed protons and electrons are jammed together to make neutrons. The collapse is abruptly halted when these and all the other neutrons begin to interact. At a radius of 10 kilometers or so, like the electrons of a white dwarf, they too turn degenerate. The stop is so violent that it presumably sends a monster shock wave (aided perhaps by a flood of neutrinos and who knows what else, perhaps violent convection, theory still far from complete) back through the outer parts of the star, including the nuclear burning shells. The tremendous heat, in the billions of degrees, causes an expanding nuclear holocaust in the old core's shells and in the outer hydrogen envelope that tries to drive everything to iron and nickel. The expansion and cooling, though, do not let things get that far. The result, along with the rapid capture of neutrons, is the creation of all the elements of the periodic table beyond hydrogen, including a tenth of a solar mass of new iron and the really heavy stuff like uranium. The optical supernova is so spectacular that it's visible across much of the Universe. And the visible part is but a hundredth of the total energy output. Most of the stuff of which Earth and we are made was created in supernova explosions long before the Sun was born 4.5 billion years ago. Proof of the concept is that we observe the optical radiation to be driven first by radioactive nickel (Ni-56), which decays into radioactive cobalt (Co-56), which in turn decays into stable iron, the Fe-56 from which we build our cities.

The great supernova of the year 1054 that produced the Crab Nebula was caused in this way as was the one that appeared in the Large Magellanic Cloud (a small companion galaxy to ours) in 1987. Though more than 150,000 light years away, it hit third magnitude and was easily visible to the naked eye. We also caught 11 neutrinos from it, just about the expected number, giving us a look inside the self-destroying star and, not least, confirming theory. For a few moments the Earth was flooded by more neutrinos from the distant supernova than we were getting from the Sun.

The explosion's debris, rich with freshly-made elements, rapidly expands outward as a "supernova remnant" (SNR). Most supernova remnants, however, are not made of the stellar ejecta but are created by energetic shock waves (Chapter 5) from the explosions that are penetrating through the local interstellar media. The shocks are so powerful that they radiate X-rays while at the same time creating extraordinarily beautiful structures up to hundreds of light years across that rival the aesthetics of planetary nebulae.

The shocks also accelerate protons, electrons, even heavy atomic nuclei to speeds near that of light. Trapped into vast orbits by our galaxy's magnetic field, cosmic rays smash into the Earth's upper atmosphere where they produce showers of subatomic particles that spray to the ground. You are sitting in them right now: see Chapter 5.

Left behind

The old core collapses to become a neutron star in which 1.5 or more solar masses are stuffed into a ball the size of Manhattan with a density that of nuclear matter, 100 trillion grams per cubic centimeter: 100 million metric tons in a sugar cube. The magnetic field is also compressed to gigantic proportions, to strengths a trillion or more times that of Earth. Conservation of angular momentum (it appears everywhere) makes the developing neutron star-to-be spin ever faster, up to many times per second. As is Earth's, the field is tilted to the rotation axis. Radiation beams sent out along the field axis wobble madly in space and, if the Earth is in the way, hit us with a blast, the neutron star now also called a "pulsar" (introduced in Chapter 4). More than a thousand are known. They don't actually "pulse": they rotate. Recent ones, like that in the Crab, which spins 30 times per second, are ultra-energetic and radiate across the spectrum, even into the gamma-ray domain. As they radiate away their energy, winds coupled with magnetic fields slow them down and they are visible (like the first ones found) only in the radio spectrum. By the time their periods have stretched to eight or so seconds, most disappear from view. Given roughly one core-collapse supernova per century and the age of the galaxy, there must be some 100 million dead neutron stars in our system. At least one is nearby. Maybe. The detonations of core-collapse supernovae seem to be off-center, which gives the resulting neutron stars high speed kicks, enough to send some them into the galaxy's halo, if not out of the galaxy altogether.

Peculiarities abound. The expected neutron star of Supernova 1987a has never been found. Neutron stars develop an outer iron crust. As they slow their spins, their structures must change, but given the solidification, they do so at times of their choosing and quite violently. At the extreme, where magnetic fields can climb to

100 times "normal," radiation from the "glitches" can disrupt the Earth's ionosphere even though the culprit be thousands of light years away. Pulsars in binary systems can accrete mass from their orbiting companions, resulting in X-ray pulses and bursts. Mass drawn from the companion first enters an accretion disk from which it falls onto the neutron star, hitting it on the side, which gradually makes the pulsar rotate ever faster, hundreds of times per second. Millisecond pulsars are the most accurate natural clocks known. By donating its matter, the companion is eaten away, eventually evaporating altogether. Planets may form from the neutron star's accretion disk, and were among the first known. Don't look for life there: the neutron star's magnetic field would be lethal.

As there is a mass limit to white dwarfs (the Chandrasekhar limit at 1.4 solar masses), so there is one to neutron stars. It's difficult to calculate, but seems to be around three times that of the Sun. Beyond that limit, the degenerate neutrons can no longer provide support. A neutron star's fate is then to collapse forever as it turns into a black hole from which light cannot escape. There are two ways of looking at black holes. At the "surface" of a black hole, which for stellar masses has a radius not far beneath that of a neutron star, the escape velocity hits the speed of light, and light can't get out. Light, however, always maintains its speed in vacuum, one of the basic tenets of relativity. It can't lose energy by slowing, as a thrown ball would do, so it does it by reddening, by increasing its wavelength. At the surface, the "event horizon," the light reddens to infinity, leaving nothing but blackness.

We are as certain as can be that supermassive black holes thrive within the hearts of galaxies (Chapter 4). The evidence for stellar-sized black holes is pretty secure as well, as there is a handful in double star systems, wherein a tidally distorted star (tides explained in Chapter 2) feeds matter into a disk around a black hole companion. The disk gets so hot it radiates X-rays. The first and best known of these is Cygnus X-1, in which the donor star is a supergiant that first has no business emitting X-rays and second is being shifted in an orbit by a massive, but invisible, body (the supergiant's motion obvious through the Doppler shift in its absorption lines). We can

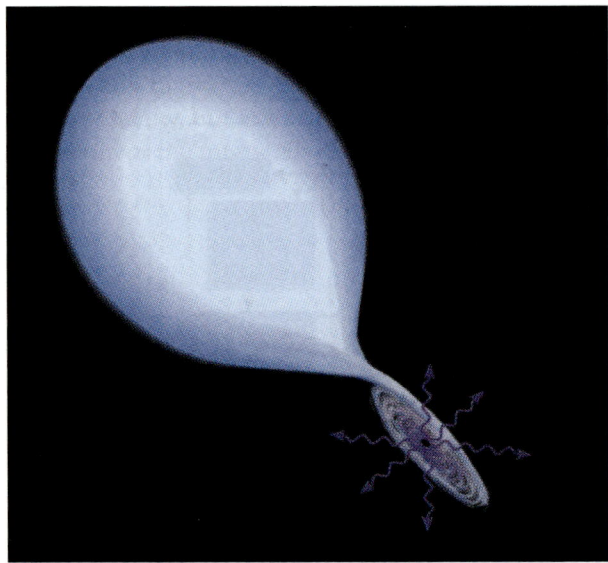

Figure 7.5. In rare double stars like Cygnus X-1, mass from a tidally distorted supergiant flows into a disk around a black hole that is the end product of the explosion of a star with a birth mass greater than that of the current supergiant. Feeding matter to the black hole, as the disk circulates it becomes so hot that it radiates X-rays. From "Astronomy! A Brief Edition," J. B. Kaler, 1997.

only conclude that the companion to the supergiant is a true black hole. Black holes may be the leavings of stars above some initial mass limit, say 40 or so Suns, but the kinds of stars that ultimately produce them is not really known.

Collapse to black holes has been related to gigantic explosions from deep space that produce bursts of gamma rays first detected by a defense satellite in the 1960s that was designed to check for Soviet nuclear blasts. Though rare within any given galaxy, there are so many galaxies (billions and billions, even trillions) that we see at least one of these gamma ray bursts ("GRBs") every day. The burst is beamed out through the rotation axis of the collapsing star and, if we are in the way, we see a high-energy flash. As the gamma rays hit the surrounding matter lost previously by the star through its wind, the GRB creates an optical afterglow, the brightest of which might be visible to the naked eye even though billions of light years away.

The best candidate in our galaxy for a GRB is the southern star Eta Carinae (Chapter 4 and Figure 4.13). Set within the bright Carina Nebula, Eta Car weighs in at a birth mass of around 100 solar masses. In the 1840s, the star, now just modestly visible to the naked eye, underwent a gigantic eruption that made it the second brightest star in the sky after Sirius, while producing an expanding dusty cloud that virtually buried the eruptor itself. Though 6000 light years away, if Eta Car went supernova, its gamma ray burst could be strong enough to affect Earth's atmosphere. Fortunately, the rotation axis seems to point somewhere else. How a close massive orbiting companion factors into the eruptions and future GRB is unknown but thought to be of crucial importance.

There are two different kinds of GRBs. Energetic supernovae are associated with bursts more than two seconds long. Short bursts constitute about a third of the total. They are probably the products of mergers between neutron stars, between neutron stars and black holes, or (as found from LIGO, the gravity wave detector) even paired black holes in binary systems. It's been suggested that a nearby GRB might have produced one of the biological mass extinctions that Earth has suffered. Even "ordinary" supernovae could do harm if within about 30 light years by damaging the ozone layer and disrupting the bottom of the food chain. And if the bottom goes out, the top is not going to be doing so well. If you are close enough to either kind, it won't matter much.

It's Ia time

Given that there are "supernovae," where are the "novae" from which the term is derived? They are common, even plentiful. Over the twentieth century we were delighted with half a dozen that reached, or at least came close to, first magnitude. Novae (from Latin meaning "new") are produced in tight double systems in which an ordinary dwarf, usually one of low mass, is close enough to be tidally distorted and thus to feed matter to a dense white dwarf. Even giants can serve as donors. When the fresh hydrogen layer is compressed and hot enough, it rips a nuclear explosion via the carbon cycle that tears off the deposited surface layer and brightens the star enormously. The system then returns to normal as it waits to produce another blast thousands of years hence.

Supernovae at first just seem to be at the bright end of the range. Their luminosities, far above those of novae, show differently. In the 1930s, Rudolph Minkowski (1895–1976) and Fritz Zwicky (1898–1974), both working at the Mt. Wilson Observatory, subdivided supernovae seen in other galaxies into Types I (with no hydrogen lines in their spectra) and II (with hydrogen). Type II supernovae, seen to be confined to the disks and spiral arms of galaxies, were rather obviously the explosions of massive stars and were later identified as being caused by core-collapse. In contrast, Type I's appeared all over the place, including galaxy halos, which are repositories of low mass stars, and in elliptical galaxies, which have no massive stars at all. Type I was later divided into Ia, Ib, and Ic. The latter two were also identified with massive stars but that had lost their outer hydrogen envelopes through winds.

That leaves the Ia's, which remain widely distributed within galaxies and cannot in any way be produced by high mass stars. By default, Ia's are explained by the overflow of mass onto a white dwarf, the same as happens to make a nova. The difference is that if the white dwarf is already fairly massive and can be pushed over Chandrasekhar limit of 1.4 Suns, it will not just have a surface eruption, but will totally collapse and flame out in a gigantic nuclear bomb, annihilating itself. An alternative possibility is the merger of double white dwarfs that together top the Chandrasekhar limit. The Ia's are generally brighter than the other kinds and make even more iron, about triple what the Type II core-collapsers offer. The most famed supernova of recent times, Tycho's Star of 1572 in Cassiopeia, which again reached the brightness of Venus, was from its behavior Type Ia, as were Kepler's Star of 1604 in Ophiuchus and the record-holder of the past two millennia, the supernova of 1006. In the southern constellation Lupus, SN 1006 was said by some to equal the brightness of a quarter Moon. All the historical supernovae have expanding gaseous remnants to commemorate their existence.

Ia supernovae also have the remarkable property of being very uniform and thus, once calibrated, make wonderful luminous standards by which we can obtain accurate distances of other far-away galaxies and therefore judge the properties of the Universe. Type II (and related) supernovae and their relatives generally have too great a range in intrinsic luminosity for them to be of much use. From the Ia supernovae we have found the expansion rate of the Universe to be increasing as a result of some kind of "dark energy" that dominates the mass-energy content of space. Summing the mass equivalent of the dark energy (74 percent of the Universe's mass); dark matter, which has a gravitational field, but radiates nothing (22 percent); and ordinary matter (4 percent), leads to the density needed to "flatten" the Universe and give it a Euclidean geometry. Stars make up a mere 0.6 percent of the Universe's stuff (the remainder of ordinary matter in the form of hot gas floating in intergalactic space). But when you go outside under a clear sky, it's not dark matter or dark energy you admire. It's the stars.

Looking back

As introduced in Chapter 6, the open clusters of our galaxy's disk (Hyades, Pleiades, etc.), which have more or less solar chemical abundances, range from new (the Double Cluster in Perseus) to about 10 billion years old, their maximum age giving us the disk's age. The globular clusters of the halo, however, are 12 to 13 billion years old, as are halo stars in general. With the exception of escapees, there are no massive halo stars: the halo's main sequence is burned off down to 0.8 or so solar masses. The globulars and halo thus came first, the halo collapsing to form the disk, aided strongly by mergers with other systems. The ancient globular clusters are also metal-poor, their iron contents running from a tenth to close to a thousandth the solar value. Looking into the dense galactic center, where star formation and recycling occur at a faster pace, we find stars that are super-metal rich, with iron contents double or more that of the Sun. Thus as the galaxy ages we see its metal content increasing thanks to giant-star winds and supernovae.

From a study of the Andromeda Galaxy (Figure 4.12) made in the 1940s at Mt. Wilson by the German astronomer Walter Baade (1893–1960), our own galaxy's disk, which includes young O and B stars, is called Population I, while the halo stars, including the globular clusters are of Population II. (Type II supernovae are thus Pop I, type Ia's Pop II. You get used to it.) Given that the Big Bang created nothing but hydrogen and helium (and a trace of lithium) there should be stars with zero metals, called Population III. It's an empty set. We can't find them. Halo stars take us all the way back to iron contents that are 1/100,000 that of the Sun. But "zero" eludes. Moreover it seems to take dust to cool the interstellar gases to the point where star formation can proceed. But with no metals, there is no dust. How then how did they form? Theory to the rescue. It seems you can make hydrogen-helium stars using molecular hydrogen as a coolant, but they will be massive. The first stars of Pop III thus blew up early as supernovae and there are none left to find. All we see of them are trace elements in somewhat older stars made by the rapid capture of neutrons, which unlike slow neutron capture runs to uranium and beyond. (Unfortunately we don't really know where rapid neutron capture actually occurs, whether within the supernova itself, the neutron star, or somewhere else.) The black holes created by the collapse of massive stars might then have been the seeds around which galaxies were built.

Though the whole idea sounds made up to avoid a serious problem, the record of elemental abundances is consistent, with core-collapse supernovae coming first, Type Ia's later. We can actually track elemental abundances with time until we get to those in the Sun, and we thus return to the beginning of our story. While our Earth may seem insignificant against this vast panorama, it is really quite special, as it is made from the distillate of billions of years of the nuclear debris of ageing stars, as are we ourselves.

8

Other Worlds

"But shall we recognize in Mars all that makes our own world so well fitted to our wants — land and water, mountain and valley, cloud and sunshine, rain and ice and snow, rivers and lakes, ocean-currents and wind-currents — without believing further in the existence, either now or in the past, or in the future, of many forms of life? Surely, if it is rashly speculative to form such an opinion respecting this charming planet, it is to speculate still more rashly to assert that Mars is not, has never been, and will never be tenanted by living creatures, or by any beings belonging to other than the lowest orders of animated existence."

R. A. Proctor
Other Worlds than Ours
New York, Lovell, Coryell, & Company, 1870.

Richard Proctor's daughter Mary (1862–1957) became the first woman science writer, communicating astronomy with excellence as did her father. She most famously authored "Stories of Starland" in 1895 for children.

Musings on the solar system

Belief in life on Mars continued well into the twentieth century. In 1877, the Italian astronomer Giovanni Schiaparelli observed a network of fine lines that crisscrossed the planet and named them canali, or "channels." Unfortunately, the word was simply translated into English as "canals". And who builds canals? Why, people! Thus began a prolonged, sometimes frustrating, sometimes ridiculous Martian mania. Martians, it was imagined, constructed the canals to transport water from the poles into dry desert regions, allowing the dark "oases" to expand and blossom with thick vegetation during the summer. The Sunday supplements even had drawings of how the Martians looked.

The Martians' cause was most taken up in the 1890s by the American businessman and astronomer Percival Lowell, who established an observatory in Arizona to study Mars and who fervently believed in the canals until his death in 1916. For decades, an argument raged over their existence. Only some astronomers could see them; others resolved the canals into patterns of dots. Under less than excellent conditions, these dots appeared as lines, producing the effect of canals. A far less glamorous explanation for the changes in the dark areas held that they are caused by wind patterns that shift dust around the globe. Belief in the canals and the possibility of life was so strong, however, that some observers claimed the dark areas to be green (probably a contrast effect against the redness of the planet); one astronomer even announced the discovery of the spectral signature of chlorophyll, a molecule involved in photosynthesis by plant life. Martian madness reached its apex in 1939 with the radio dramatization of H. G. Wells's story "The War of the Worlds." Orson Welles (no relation) convinced millions of listeners they were actually listening to a news broadcast of a Martian invasion. The best telescopic view of Mars from Earth, however, was at that time no better than that of the Moon with the naked eye. The discovery of the real planet, in truth more interesting than anyone imagined, had to wait for the close-up views made possible by the space age. There were no canals, only vast plains filled with dunes, craters, and immense volcanoes. Still the search, with multiple landers, goes on, with fascinating results, but with no discovery of life. Yet even then there was the "face on Mars," a shadowed rock was purported to be a sculpture by a Martian artist. It even made its way onto a postage stamp. Finally, with no life to be seen, there had to be a NASA conspiracy to hide the images of cities…

The episode of the Martian canals teaches a lesson on how science should, and usually does, operate. The essential assumption of science is that nature is not capricious; experiments or observations must be repeatable. Only Lowell and a few others could "see" the canals; those who could not were told that their equipment or their eyes were inferior. The French astronomer E. M. Antoniadi, who considered the canals false apparitions, used some of the best telescopes available in his day and could not see them. Lowell strained credulity when he attempted to explain why larger telescopes were less able to detect delicate features than smaller ones. This kind of explanation is usually a clue that something is amiss. The history of science is replete with parallel instances, one more recent. The Sun is powered by thermonuclear fusion, which creates heavy elements from light ones with the release of vast amounts of energy. On Earth, the process has been used to produce hydrogen-bomb explosions. Scientists have tried with limited success to generate nonexplosive fusion for power production, because it requires enormously high temperatures. In 1989, a pair of physicists working at the University of Utah apparently discovered a way of fusing elements at room temperature using a catalyst [a substance that aids a reaction without being changed by it]. They immediately announced the results of their "cold fusion" experiments, and scientists everywhere took note. If true, here was a potential source of cheap, reliable power. Unfortunately, few scientists could repeat the fusion reaction. So many failed, in fact, that it became clear that cold fusion does not exist. The moral of these tales is that nature is reliable. The scientific process, which depends on experimental repeatability, operates to check phenomena that are observed only by particular fortunate or "sensitive" individuals. Scientists cannot become part of the science itself. J. B. Kaler, adapted from "Astronomy!," HarperCollins, 1994.

Yet the search for life in our Solar System goes on. We can pretty much rule out Mercury and Venus. Both are too hot, while Venus is also toxic with its extraordinary carbon dioxide atmosphere. Which seriously leaves Mars. It looked for a time as if one of the Viking landers of the mid-1970s had found activity in the "soil," properly the "regolith" since there is no organic matter, but then that is what we are trying to find. Sadly perhaps, it was of chemical origin. Then there is the famed ALH 84001, a Martian meteorite found in the Alan Hills region of Antarctica in 1984. (Though the vast majority of meteorites come from the asteroid belt, a rare few have been blasted off the Moon and Mars by huge impacts. They can be identified by their chemistry. We probably have pieces of Venus and Mercury on our planet too, but no case has ever been made for any such rocks.) The ancient Martian rock at first was announced with great excitement as containing nanofossils. Life at last. Unfortunately they turned out to be (broadening the term to the planets) of geological origin. Moreover, there is the repeatability issue to deal with, as ALH 84001 was a singular discovery. So in spite of vast efforts there is still nothing. Mars may just have cooled too fast for life to take hold. Who knows? Still, the lack of life (if that is the case) is almost equally intriguing, as it ultimately tells us something about the possibly unique origin of life on Earth.

The broad assumptions are that to harbor life we need warmth (though not too much) and that mighty solvent, water. Two distant bodies stand tall. Jupiter's satellite Europa (the second large one out) is about the size of our Moon. Gravity measures reveal what may be a deep ocean beneath the ice-covered surface seen by the indominatable Voyager spacecraft and with "Galileo," which orbited the planet from 1995 to 2003. Europa is close enough to Jupiter to be strongly affected by Jovian tides. Two other satellites, Io and Ganymede, are in resonant orbit with Europa (their periods in simple ratios) and pull it back and forth, slightly changing its distance from the Jupiter and thus changing the size of the tidal bulge. The constant squeezing, ("tidal flexing") heats Europa's interior, presumably warming the deep sea. (The process reaches its apex at the inner large satellite Io, which is so hot inside that it spews sulfur-rich volcanoes and where life has no chance at all.) Life is widely thought to have originated in our seas. The idea of life on Europa is intriguing enough that planetary scientists are at least planning a spacecraft for a special visit to study it. One might also make the same argument about Ganymede, the third out. Number four, Callisto, seems too cold.

Farther out in the Solar System, we might consider Saturn. One of its satellites, Enceladus, also tidally heated, blows off water-rich plumes and displays evidence for a subsurface ocean. Though a mere 500 kilometers wide, and acting more like a comet than a moon, could Enceladus also harbor some form of primitive life? We might also question the basic assumption that water is required. Saturn's huge satellite Titan, about the size of Mercury and the second largest moon in the Solar System after Jupiter's Ganymede, has what appears to be a full hydrologic cycle based on methane (NH_3), that includes clouds, rain, and lakes. We parochially assume that "everybody" (if there is anybody) must be like us, a continuing theme. But are they? We have no idea. For that matter, we might make a case for life on just any body of the Solar System at all. But it sure is not jumping out at us. So we have to look outside, to use Proctor's book title, at "other worlds than ours." But before speculating on life on other worlds, we have first to find the other worlds themselves.

Exoplanets
(Those belonging not to the Sun but to other stars)

The problem of discovery of any kind is often just a matter of technology. Bacteria were unknown until the invention of the microscope, Solar System satellites until that of the telescope. Brown dwarfs, though theorized, were only marginally known until the advent of infrared surveys, and then they literally fell out of the sky. Brown dwarfs are everywhere.

Planets orbiting other stars suffered the same indignity. They were found and then unfound, discovery withdrawn or disbelieved. The most infamous case involved Barnard's Star, a tenth magnitude red dwarf only six light years away (the next-closest star after the triple system Alpha Centauri) that holds the record for proper motion, skimming along against the distant background at a rate of 10 seconds of arc per year (highlighted in Chapter 6), an angular movement easily detected with an amateur's telescope. In spite of its faintness, Barnard's Star, named after its discoverer, E. E. Barnard (1857–1923) of Yerkes Observatory, assumed the mantle of greatness when decades-long observations showed it not just to arrow forward, but also to wobble sideways back and forth just enough for an astronomer at a well-known eastern observatory to suspect an orbiting planet half again the mass of Jupiter. The same technique was used in the initial discovery of Sirius B: see Chapter 6. Alas the mantle was shredded when the wobble was found to be the result of the repositioning of the telescope lens during maintenance. Moreover nobody else could confirm the planet (see the canals of Mars). No exoplanet seems to attend Barnard's Star. Which doesn't mean that there is not one there; one can't prove a negative in this business. Epsilon Eridani, a K dwarf 10.5 light years away, suffered a similar fate, its proper-motion wobbles spurious, though in this case it was later found to have a real Jupiter-like planet in a seven-year orbit. There are many more similar examples.

For indirect evidence for exoplanets, just look out at our own planetary system. It seems to have had a natural origin within a flattened disk surrounding the early Sun (recall Kant and Laplace), hence planets. Then zero in on Jupiter, even Saturn. Their satellite systems appear to be miniature orbiting "solar systems," the planets now acting as little suns. Moreover, why should we be unique? The whole history of astronomy has displaced us ever farther from centrality. Should not all stars have planets attending them? These speculations were finally realized, but only through sufficiently advanced technologies. When applied, planets, like brown dwarfs, again fell out of the sky. How do we find them? As Elizabeth Barrett Browning (1806–1861) almost wrote, "Let us count the ways."

"Finding Nemo"

It wasn't easy. One might think that all one had to do to see planets orbiting other stars would be (see Yogi, Chapter 3) just to LOOK.

One would be wrong, as other planets, at least those like ours, would be tucked up against their glaring stars and be overwhelmed. If you were on Alpha Centauri, the nearest star, Jupiter would be a mere 25th or so magnitude dot just a few seconds of arc from a zeroth-magnitude star (our Sun) in the Milky Way in eastern Cassiopeia. (From only four light years away, the constellations would look pretty much the same.) Actually the technique does work (see far below), but with difficulty; other ways are currently much more effective.

The first real inkling of other planetary systems did not involve the planets themselves, but the discovery of warm circumstellar disks reminiscent of that around the Sun caused by comet dust and debris from asteroid collisions. Made visible by reflected sunlight, the "zodiacal light" can be seen from Earth as a faint luminous cone stretching upward from the horizon through the constellations of the Zodiac along the ecliptic after the end of evening twilight or before morning's light, when it is called the "false dawn." From the northern hemisphere, it's especially noticeable in fall evenings and spring mornings when the ecliptic is most vertical to the horizon. Add to that the Gegenschein, or counterglow, a dim patch of light directly opposite the Sun caused by backward reflection from interplanetary dust grains. When the sky is especially dark during solar magnetic minimum, with minimum airglow from impacting solar particles, the zodiacal light can be seen crossing the heavens like a faint tilted version of the Milky Way.

Observing the infrared sky for nearly a year in 1983, the Infrared Astronomical Satellite (IRAS) discovered a number of class A stars radiating excessively at long wavelengths, implying a similar warm debris disk, the list including Vega, Fomalhaut, Denebola (Beta Leonis), and perhaps most famously, Beta Pictoris. Where there is debris, there must be something to make it or to come from it, namely planets and their attendants. Beta Pic was eventually seen to be buried in a huge edge-on disk 10 times the extent of our own system. But circumstellar disks (which abound) are circumstantial evidence; we needed more.

The false planet orbiting Barnard's Star was found through the star's apparent side-to-side wobbling as it proceeded along its path.

But as it orbits, a planet will also make its star wobble back and forth along the line of sight, giving it a variable radial velocity (Chapter 6). The wobble is so small, measured in meters per second, that it was undetectable with traditional spectroscopy. Finally, in the 1990s, by observing stars through a gas-filled cell that provided a precise wavelength reference, astronomers succeeded. The first planet to be discovered belongs to the sunlike G2 subgiant (more likely dwarf) 51 Pegasi. With a tight orbital radius of just 0.05 AU and a period of a mere 4.4 days, the planet shifts the star through a maximum speed variation of about 50 m/s in each direction. While the Doppler technique gives the orbital period and radius (assuming the planet is miniscule compared to the star), even the orbital eccentricity, it provides only a lower limit to the planet's mass. Nevertheless, 51 Peg's planet carries more than half Jupiter's mass. Other similar planets followed.

Now wait a minute. See Chapter 5. Jupiter was supposed to have formed out beyond the "snowline" where the early solar nebula around the developing Sun was cold enough for the planet to accrete volatiles like hydrogen, helium, water, etc., so much that it could grow large. What is "Jupiter" doing in a short orbit 0.05 AU from its star? Most likely, 51 Peg's planet did not form there, but was born way out and migrated inward as a result of turbulent and gravitational "friction" with its forming disk, which dissipated just in time to prevent the planet from actually falling into its star. The speculation is quite reasonable actually, as there is strong evidence that our own Jupiter severely migrated inward after formation, the other planets moving around as well before taking up their final places. In the early days of exoplanet discovery, hot jupiters were all over the place, common, their seeming number a selection effect as they cause the greatest and most easily observed radial velocity variations. Given longer time baselines and improved abilities, with uncertainties in measurement reduced to less than a meter per second, more planets, including those with much lower masses, even whole planetary systems, were revealed. If the planetary orbits are face-on, however, there will be no observed velocity variations. Vega seems to fall in that class.

Another productive technique uses planetary transits. Planetary orbits will naturally have all sorts of inclinations to the line of sight. In a small fraction of cases, the planet crosses in front of its star, producing an "eclipse," the brightness of the star dropping by a tiny amount. For the method to work, the technology of stellar magnitude measure had to be improved to sub-millimagnitude precision. While planetary transits are observed in some abundance from the ground, the best place to find them is from space, where there is no turbulent air to jiggle stellar brightnesses. Launched in 2009, the Kepler mission continuously monitored around half a million main sequence stars from class F through M in a 12-degree-wide field of view near the Cygnus-Lyra border over a period of several years and discovered more than a thousand planets and planetary systems, with thousands more awaiting confirmation. A major problem involves the variation of the star itself, which can at first mimic a transit, so more than one event is required for confirmation. Kepler then turned the problem around to study stellar oscillations, which can be used like terrestrial seismography to examine stellar interiors. While the transit technique can measure planetary diameters and orbital periods, it cannot derive masses or orbital sizes. Combine it with Doppler measures, however, and all the parameters of the system are revealed. The planet's radius combined with mass then yields the density. Eventually we will almost certainly find an "earth."

And indeed, there is also direct imaging. The corona of the Sun can be observed from the ground at high altitudes by blocking out the bright Sun with an occulting disk internal to the telescope. Before the space age rendered them less useful, coronagraphs were in operation in the high Rocky Mountains and French Pyrenees. Stellar coronagraphs that block out starlight view not the stellar coronae but orbiting planets. The first to be found were planets attached to the first magnitude star Fomalhaut (in Piscis Austrinus), Beta Pictoris, and right next to 51 Pegasi, HR 8799, all of which are class A stars that have surrounding debris disks, so planets were no surprise. Yet after waiting decades, if not centuries, just seeing them is amazing and deeply satisfying, though discerning their characteristics is daunting.

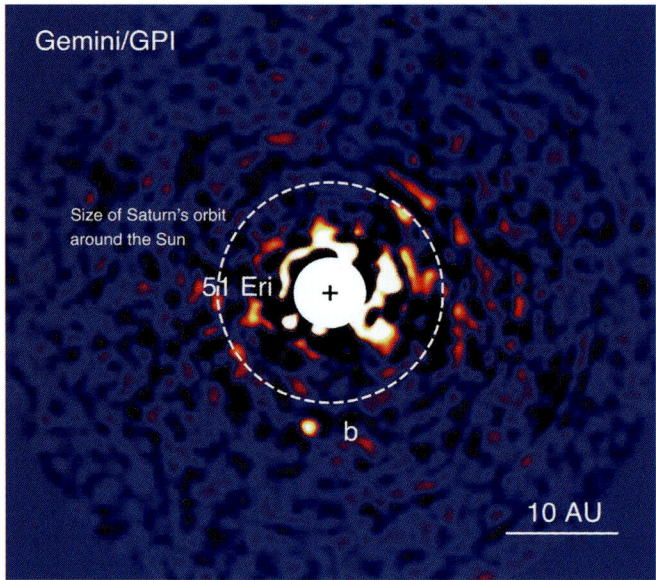

Figure 8.1. With the light from the star 51 Eridani mostly blocked out by an occulting disk in the telescope, a planet similar perhaps to Jupiter, 51 Eridani b, emerges from the darkness, its orbital radius greater than that of Saturn. Thousands of extra-solar planets have been found by various means. Nobody has yet been in touch. J. Rameau (U. of Montreal) and C. Marois (National Research Council of Canada, Herzberg).

A fourth successful technique uses gravitational lensing. Through the mathematics of relativity, gravity is viewed as a distortion of spacetime produced by any sort of mass (Chapter 2). As a ray of starlight passes close to the Sun, it bends. Measurements of the displacements of stars during a solar eclipse was the first proof that the theory was, and is, correct. If a source of light is placed directly in back of a much closer mass, it will be magnified into a brighter ring or, if offset, into a variety of shapes. The phenomenon was first seen in the distortion of distant quasars as their light traversed a closer galaxy. (Quasars, or QSOs, "quasi-stellar objects," are luminous, accreting black-hole nuclei of developing galaxies in the early distant Universe. They were first "seen" as radio sources, hence the original term "quasar," which stands for "Quasi-Stellar Radio Source". Most are actually radio-quiet.) Cosmologists map dark matter, which has the gravitational pull of normal matter but is invisible, in nearby clusters of galaxies through the relativistic distortions of distant background galaxies. Closer to home, planet-hunters simultaneously monitor millions of stars (made possible by digital detectors and massive computation) to find one that suddenly brightens as a planet passes in front of it. While the Doppler technique works well for close-in planets, lensing complements it for planets distant from their suns, as does direct imaging, though it's difficult to apply and yields few results.

Once a planet is discovered orbiting a star, variations in its smooth sinusoidal velocity curve (the graph of stellar velocity against time) can be used to infer yet another planet, even two, three, or more, giving us whole new "solar systems." Even moons orbiting their planets can potentially be located this way. The transit system is ideal for discovering planetary systems. Since planets are distributed in relatively flat planes, if you find one planet by transit, the odds are high that you will find another. Using all the techniques, we've worked our way down in mass to "superearths." Some stars seem planetless (or their planets remain just undiscovered), while others have systems up to eight orbiting bodies. Our old parochial view was that once found, all planetary systems would be more or less just like ours. In spite of locating enormous numbers of planets and planetary systems, we have yet to find our clone. It does seem though that the majority of stars have planetary systems (however they are constructed) and that planet formation is indeed a natural part of star formation. Planets seem everywhere. Even "free floaters" have been found, perhaps planets that have been gravitationally kicked out of their systems, or even formed directly as low mass objects that were never intended to be planets in the first place, just ultra-low mass brown dwarfs. In spite of stunning advances, we are still a long way from a complete, even partial, inventory of celestial objects. Indeed, we have yet to have a complete inventory even within our own Solar System.

But is anybody there?

The goal, obvious even if unstated, is to find life. A first step would be to locate an earthlike body within a "habitable zone," defined as the range in distance from a star over which one might find liquid water (perhaps another parochial view). In our Solar System the zone is pretty small. Venus is a raging atmospheric furnace, Mars a freezer (though one wonders what it would be like with greater mass, enough to have trapped a significant atmosphere). The size of a habitable zone and the distance from its star depends on the stellar luminosity, the distribution of its radiant energy (infrared, visual, ultraviolet), the nature of the planetary atmospheres, their masses, and other variables.

Red dwarfs are particularly interesting targets as there are so many of them and their low masses respond especially well to planetary gravitational pulls. On the downside, a good fraction of red dwarfs are magnetically active "flare stars" with magnetically-induced sunlike flares encompassing the stars and brightening them by factors of two, three, or more, especially in the ultraviolet or X-ray domains. Try sunbathing in that. Moreover, the habitable zone of a low-luminosity red dwarf must necessarily be very small and close to the star. As a result, the "year" will be short, the question of "Whatever happened to summer?" taking on a whole new meaning. Moreover, a planet within the red dwarf's wet zone will likely be in synchronous rotation (as the Moon is to the Earth), with one side facing the star at all times, the other facing away, yielding strong heating and extreme radiative cooling unless there is a thick atmosphere to distribute heat.

The next great step is to get good spectra of the atmospheres of extrasolar planets, which can be done when planets' atmospheric blankets are seen during eclipse or even done directly in the case of imaged planets. Methane would be a signature of life on an earthlike planet since the stuff is constantly destroyed by light from the central sun and must be renewed, and what better way for it than through life processes? Loaded with methane, Saturn's Titan perhaps tells us differently. But except in "Farside" cartoons, bacteria, grass, and cows don't talk. The real search is for INTELLIGENT life. This quest has nothing to do with UFOs, "Unidentified Flying Objects." For decades, people have seen funny lights in the sky, or thought they had. Since nobody knew what they actually were, why not identify them as being crafts driven by "space aliens?" That sure makes sense. UFOs are a wildly mixed bag. Venus tops the list. So bright that it can throw shadows at night and be seen in the daytime sky, it follows along with you as you walk or drive. People have called the police on it, for that matter the police have even chased it. Then there are Jupiter, Mars, meteors (comet fluff or small asteroidal debris that burns up upon hitting Earth's atmosphere), sundogs and subsuns (Chapter 1), odd and experimental aircraft, hoaxes, conspiracy theories, and delusions. A conjunction of Jupiter and Venus a few years back was called into the cops several times. Nobody has ever encountered an alien, and nobody has any proof they exist. No, there is no alien body in Area 51. Amateur and professional astronomers, who make a business out of watching the night sky, never see UFOs. (Well, hardly ever. Your author saw one once when he was 14 or so while observing with his wobbly 3-inch telescope. It was a bright crescent-shaped light that moved quickly across the sky. He still doesn't know what it was. So it's "unidentified.")

To find out if intelligent life really exists, we need communication that reaches out over the light years. We make the somewhat rash assumption that another civilization will use radio signals, as do we. The first person to look for them was Frank Drake, who used the 150-foot radio telescope at the National Radio Astronomy Observatory in West Virginia. Picking a couple nearby sunlike stars, Epsilon Eridani and Tau Ceti (respectively 10.5 and 11.9 light years away), he found nothing, though a singular early positive gave some hope that was dashed when it never came back. His work has morphed into various guises, chief of which is an institution dedicated to the "Search for Extraterrestrial Intelligence," now the "SETI Institute," which has the largest detection system in the world, one that uses a specialized radio telescope. After examining countless stars, there is still nothing, no obvious signal, no modulated AM or FM, no code.

But what are the odds of finding anything? They are expressed through the "Drake Equation," which involves the numbers of planets per star, the fractions of those in habitable zones, of those that go on to support not just life but intelligent life then communicating life. It also factors in the length of viability of the civilization. How long does one last? About the only thing we have any handle on are the commonality of planets and the fractions of these in habitable zones. Other small fractions are expected to be offset by the sheer numbers of stars, 200 billion or more in our galaxy alone.

Are other civilizations actively trying to contact us? If so, at what wavelengths do we look? There is a naturally-quiet band of the radio spectrum between the interstellar 21-centimeter emission line of neutral hydrogen and interstellar water vapor called the "water hole." Anyone capable of sending a deliberate signal would surely know of it. Then what would they send? We surely would not understand their language or image formats, if they have any at all. Perhaps "pi" in binary to let us at least know they are there? And then there is the matter of funding for such projects, which is now mostly in private hands, government research offices shying away from them.

Of course we can also look for weaker signals aliens might broadcast for their own use that escape their atmospheric ionospheres. We've been broadcasting powerful radio signals into the cosmos since the 1920s. There is now a bubble of radio radiation expanding around us almost 200 light years across that encompasses more than a million stars. Our planet is "radio bright," and anyone seeking us will find us, the intensity and wavelengths of radio radiation varying periodically as our planet rotates. Imagine their view of us by watching our television shows, from Howdy Doody (you have to be of an age) to the Next Big Thing on broadcast TV, assuming of course that a civilization actually continues to broadcast. Maybe they use lasers or something else that we don't know about yet. Our ignorance is stunning. Maybe we just need better technologies to make civilizations "fall from the skies" like brown dwarfs and planets. Then again maybe it would not be such a good thing to announce ourselves. Fortunately distance seems to make space travel impossible. Nevertheless, it's been suggested that we do not deliberately tell we are here, though it's hard to believe that anything could happen that is worse than what we do to ourselves.

Maybe too we are the only ones there are. After all it took the whole galaxy to make us, to build up the chemicals out of which we are made. Maybe we are the forerunners of what will someday become an "intelligent galaxy." But even if there is but one civilization per galaxy, there are a trillion galaxies out there and those are just the ones we can, or could, see. There might then be trillions of those like us. Even if we find somebody in our own galaxy, it would be a slow conversation, each communication taking hundreds, thousands, of years, depending on how far the planet is from us. And when we look to other very distant galaxies, we see them as they were millions, even billions, of years ago, long before their civilizations had developed.

Yet for all our ignorance, we have at least learned that we are the children not just of Earth, or even the stars, but of the entire Universe. It took all of it to make us. Far from being insignificant motes on a random planet, we are not only the progeny of the whole affair but can look outward to understand and appreciate at least a little bit of the grandeur of what nature has given us. And if the past is an instructor, there are many more surprises yet to come as we go from the Sun to the stars, and then back again to our own home. Richard and Mary Proctor would have loved it.

Index